Radioactive Waste: Politics, Technology, and Risk

UNION OF CONCERNED SCIENTISTS

The Union of Concerned Scientists is a Cambridge, Massachusetts-based nonprofit coalition of scientists, engineers, and other professionals concerned about the impact of advanced technology on society. UCS has conducted independent technical studies on a wide range of questions relating to the strategic arms race, air and water pollution, unrestricted pesticide use, liquefied natural gas transport and storage hazards, nuclear power plant safety, radioactive waste disposal options, and energy policy alternatives.

Radioactive Waste: Politics, Technology, and Risk

Ronnie D. Lipschutz
Union of Concerned Scientists

A Report of the Union of Concerned Scientists

Ballinger Publishing Company • Cambridge, Massachusetts
A Subsidiary of Harper & Row, Publishers, Inc.

This book is printed on recycled paper.

International Standard Book Number: ISBN: 0-88410-621-7

Library of Congress Catalog Card Number: 79-19649

Printed in the United States of America

Library of Congress Cataloging in Publication Data

Lipschutz, Ronnie D
    Radioactive Waste: Politics, Technology and Risks

    Bibliography: p. 225
    1. Radioactive waste disposal. I. Title.
TD898.L56   614.7'6   79-19649
ISBN 0-88410-621-7

To Nan

※

Contents

List of Tables   xi

List of Figures   xiii

Acknowledgments   xvii

Introduction   1

Chapter 1
The Nature and Hazards of Radioactivity   5

The Nature of Radioactivity   6
Radiation Exposure ⁻   13
Pathways in the Environment   18
Radiation Exposure Standards and Guidelines   19

Chapter 2
The Nuclear Fuel Cycle: How Radioactive Wastes
Are Produced   29

Radioactivity and Radioactive Wastes   33
Generation of Radioactive Wastes   34

Chapter 3
The Management, Storage, and Disposal of
Radioactive Wastes   55

Packaging Radioactive Wastes: Matrixes and Canisters   56
Waste Storage and Disposal Technologies   62
The Hazards and Risks of Radioactive Waste Management   96

Chapter 4
The History of Radioactive Waste Management   113

High Level Waste   114
Reprocessing   122
Low Level Waste   125
The Mill Tailings Problem   135

Chapter 5
The Present Radioactive Waste Management
Program   139

The Current Federal Waste Disposal Program   140
The Human Problems   155

Chapter 6
Requirements for a Successful Program—
Conclusions and Recommendations   161

Reviewing the Problem   162
Requirements for a Successful Program   168
Recommendations   171

Appendixes   175

A — Radionuclides in Radioactive Waste   177
B — Health Risks Associated with Failed Radioactive
        Waste Management Facilities   187
C — Foreign Programs in Radioactive Waste Management   195

Glossary 203

References 215

Bibliography 225

Index 241

About the Author 247

✳

List of Tables

1-1   The Uranium-238 to Lead-206 Decay Chain   10

1-2   Sources of Radiation Exposure in the United States,
       1978   14

1-3   Biological Effects of Ionizing Radiation   16

1-4   Radiation Dose Limits Recommended by the National
       Council on Radiation Protection and Measurements,
       1971   21

1-5   Maximum Permissible Concentrations for Several
       Radionuclides Found in Spent Reactor Fuel   24

1-6   Maximum Permissible Concentration in Milk of
       Several Radionuclides   25

1-7   Typical Dose Conversion Factors for Exposure to
       Selected Radionuclides   26

2-1   Radioactive Waste Generated Annually in the Fuel
       Cycle of a 1,000 Megawatt (Electric) Light Water
       Reactor with a 30 Metric Ton Fuel Off-load   32

2-2   Radioactive Inventory of One Metric Ton of Spent
       Reactor Fuel   43

2-3   Thermal Output of One Metric Ton of Spent Reactor
       Fuel   44

2-4   High Level Waste and Spent Fuel Storage in the
       United States   48

3-1   Ingestion Hazard Index of High Level Defense Wastes
        in Storage as of 1976   100

3-2   Ingestion Hazard Index of Spent Fuel in Temporary
        Storage as of December 1979   102

3-3   Volumes of Water on Earth and in Several Lakes
        and Rivers   104

4-1   Low Level Waste Inventories in the United States
        as of January 1, 1977   126

A-1   Radionuclides in Spent Reactor Fuel   178

A-2   Specific Activities of Important Nuclides in
        Radioactive Waste   180

A-3   Radionuclides in Spent Reactor Fuel with Half-
        lives Shorter than Ten Years   182

A-4   Radionuclides in Spent Reactor Fuel with Half-
        lives Between Ten and Twenty Years   183

A-5   Radionuclides in Spent Reactor Fuel with Half-
        lives Between Twenty and One Thousand Years   184

A-6   Radionuclides in Spent Reactor Fuel with Half-
        lives Between 1,000 and 100,000 Years   185

A-7   Radionuclides in Spent Reactor Fuel with Half-
        lives Greater than 100,000 Years   186

✳

List of Figures

1-1   Exponential Decay Curve for Cesium–137, a Radio-
      nuclide with a Thirty Year Half-life   12

1-2   Possible Dose-Response Curves for Exposure to
      Low Level Radiation   23

2-1   The Nuclear Fuel Cycle Without Reprocessing of Spent
      Reactor Fuel   30

2-2   The Nuclear Fuel Cycle with Reprocessing   31

2-3   Slightly Enriched Uranium Dioxide Fuel Pellets Used
      in a Light Water Reactor   37

2-4   Fuel Loading at Baltimore Gas and Electric's Calvert
      Cliffs Reactor   39

2-5   Total Radioactivity for Fission Products and Transuran-
      ics Contained in One PWR Spent Fuel Assembly as a
      Function of Decay Time   40

2-6   Thermal Power Released by the Radioactivity of High
      Level Waste from Reprocessing 1 Metric Ton of
      Irradiated LWR Fuel   41

2-7   Spent Reactor Fuel from the High Neutron Flux
      Experimental Reactor   42

2-8   High Level Radioactive Waste Storage Tanks Under
      Construction at the Savannah River, South Carolina,
      Defense Reprocessing Facility   47

2-9   High Level Waste and Spent Fuel Storage Sites
       in the United States   49

2-10   Estimated Civilian and Military High Level Waste
       Inventories Measured in Terms of Their Strontium-90
       Content as a Function of Time   51

3-1   Borosilicate Glass Containing "Synthetic" High Level
       Radioactive Waste   58

3-2   Spent Fuel Storage at General Electric's Nonoperational
       Reprocessing Facility in Morris, Illinois   64

3-3   Rock-melting Disposal of High Level Liquid Wastes   67

3-4   Artist's Conception of a Geologic Repository and its
       Support Facilities   69

3-5   Test Heaters Emplaced in the Floor of the Lyons,
       Kansas, Salt Mine During Project Salt Vault   70

3-6   Bedded and Dome Salt Deposits in the United States   71

3-7   Granitic Formations in the United States   72

3-8   Basalt Formations in the United States   73

3-9   Shale Formations in the United States   74

3-10   Subseabed Emplacement Concepts   86

3-11   Space Disposal Solar Orbit Ejection Sequence   89

3-12   Ice Disposal Concepts   93

3-13   Ingestion Hazard Index of High Level Waste and Spent
       Fuel from a Light Water Reactor   98

3-14   Ingestion Hazard Index per Metric Ton of Spent Fuel
       Discharged from a Light Water Reactor Compared
       to a Typical Uranium Ore   99

4-1   Typical Storage Tank for Low Heat (nonboiling) High
       Level Radioactive Wastes   115

4-2   Solidified Salt Cake Inside a Hanford Waste Tank   117

4-3   The Retrievable Surface Storage Facility   121

4-4   Low Level Waste Burial Sites in the United States   128

4-5   Typical Crib for Disposal of Low Level Liquid Wastes   130

4-6   The Movement of Radionuclides Beneath a Typical
       Disposal Crib   131

4-7 Disposal of Low Level Radioactive Wastes in a Shallow Burial Trench at the Maxey Flats, Kentucky, Disposal Site 133

4-8 Abandoned Mill Tailings Pile in New Mexico's Uranium Belt 136

5-1 Artist's Conception of the Waste Isolation Pilot Plant in New Mexico 146

5-2 Diagrammatic Illustration of a Composite Breccia Pipe 148

5-3 Topographic Map of one of the Subsurface Layers at the WIPP Site 150

5-4 Topographic Map of the Subsurface Layer Just Below the Salado Salt Bed 151

5-5 Oil and Gas Leases Within the WIPP Repository Site 153

5-6 Potash Leases Within the WIPP Repository Site 154

6-1 A Comparison of Recent Nuclear Power Growth Estimates 164

6-2 Cumulative Amount of Spent Reactor Fuel that will be in Storage Between the Years 1976 and 2000 165

A-1 Half-lives versus Specific Activities for Selected Radionuclides 181

C-1 Schematic of Intermediate Level Waste Disposal in the Asse Salt Mine 197

C-2 Photograph of Intermediate Level Waste Disposal in the Asse Salt Mine 198

✳

Acknowledgments

The list of individuals and organizations that contributed in some way to the preparation of this book is probably much lengthier than I am able to recollect, but I am grateful for everyone's help, including those that may inadvertently be omitted.

I am particularly indebted to the Union of Concerned Scientists and the 85,000 national UCS donor sponsors for funding and sponsoring this work and to Henry W. Kendall and Daniel F. Ford for their indefatigable efforts at reviewing and correcting its contents.

I greatly appreciate the comments, criticisms, and suggestions of those individuals who reviewed the report, including John T. Edsall, Thomas C. Hollocher, James L. Fay, and Ernest J. Moniz. Thanks also go to James L. Cubie, David Jhirad, Peter Franchot, and Steven J. Nadis for their comments and advice. The author and the Union of Concerned Scientists, however, accept full responsibility for the approach, analysis, and conclusions presented here.

I am most grateful to Diane Johansen, who prepared earlier drafts of the report, and Caterina Gentile, who managed to do a better and faster job of typing the final manuscript than I ever dreamed was possible.

Finally, none of this would have come to pass without the love and sustenance of Nan Kumpf-Crownover, whose unfailing encouragement kept me going even when things got rough.

Ronnie D. Lipschutz

Cambridge, Massachusetts
July 6, 1979

✳

Introduction

Of the many issues that fuel the nuclear debate, none appears more unsettling and of so much concern to the public as the problem of radioactive waste management. Nuclear reactors produce large quantities of radioactive waste of great toxicity and persistence. The radioactive emissions from this waste are invisible, odorless, tasteless. They cannot be felt or heard. Yet minute amounts of radioactivity are capable of inducing cancer in the living, birth defects in the unborn, and mutagenic effects in the descendents of those exposed. The longevity of the radioactivity is of concern, too, for if the wastes are improperly managed, some very long-lived radioactive species may pose a continuing hazard to living things for tens, perhaps hundreds, of millenia—periods far longer than the span of recorded human history.

The nuclear power and weapons programs of the United States have generated immense quantities of radioactive waste, and there is the prospect, in a continuing program, of much more. Even so, there is not yet a demonstrated means for ensuring the safe, long-term isolation of these wastes. The programs set up by the federal government to identify and develop a means for managing and disposing of these wastes have been deeply flawed. Since its inception in the early 1940s, the radioactive waste management program has been marred by numerous accidental releases of radioactive materials into the environment, coupled with irresponsibility, false claims, and carelessness. The public has legitimate cause for concern. The diminishing credibility of the nuclear industry and the federal government on this issue has made radioactive waste disposal a major obstacle to the continued production of nuclear-generated electricity.

The paramount requirement of any successful radioactive waste management program is the protection of the health and safety of this and future generations. Many schemes have been proposed to effect the long-term isolation of the waste necessary to fulfill this requirement. Research into the more promising proposals is currently underway, but none of these disposal technologies will come to fruition before 1995 and perhaps not even during this century. Until then, radioactive wastes will continue to accumulate. They have already piled up in considerable quantities—10 million cubic feet of highly radioactive liquids, 4,000 tons of uranium fuel discharged from power reactors, 140 million tons of uranium mill tailings, 65 million cubic feet of contaminated garbage (IRG, 1978: App. D). Until an adequate disposal technology is developed, the wastes will continue to be stored, as they have been for the past thirty-five years, in near surface storage tanks, spent fuel pools, barrels, boxes, trenches, and uncovered piles—all temporary and prone to failure. Many facilities and containers have already failed. That we are now dependent on inadequate and unreliable means of storage strongly underlines the need for timely development of a permanent waste disposal technology, for the hazards of extended storage of intensely radioactive materials on the earth's surface are great. There the wastes are subject to actions of both nature and man that could cause their widespread dispersion. Indefinite dependence upon surface storage could require human guardianship for hundreds and thousands of years. There is no guarantee that this can be accomplished.

Numerous technical uncertainties remain to be resolved before a specific waste disposal technology can be implemented with confidence. There also exists a formidable array of social, economic, and political problems requiring attention. Resolution of these problems will be necessary to ensure the success of the chosen disposal technology. Failure to address these problems will, as has been the case in the past, inevitably lead to a failure of the program. Unfortunately, if the federal government's waste management program is implemented as currently conceived, it is likely to fall victim to many of the same shortcomings that have previously bedeviled the program. Furthermore, there has been and continues to be inadequate public awareness of and involvement in this nation's waste management program. As a result, government and industry have responded primarily to economic concerns rather than to concerns for the public interest and safety. Only the watchfulnesss of an alert and informed public can assure that the safety of this and future generations is considered in the formulation of a radioactive waste disposal program.

This book by the Union of Concerned Scientists provides the information about both the technical and nontechnical issues that is required for meaningful public involvement in this major societal issue. It discusses the nature and hazards of radioactivity, describes the categories and production of radioactive waste, and covers the management and disposal of radioactive waste. The report then recounts the history of the U.S. radioactive waste management program, discusses the present program of the federal government, and outlines some of the technical and societal obstacles to successful implementation of the program. Finally, conclusions and recommendations are made that, it is hoped, will offer insight into the requirements of a successful radioactive waste disposal program.

Chapter 1

The Nature and Hazards
of Radioactivity

A report on the problem of radioactive waste management must begin with a discussion of radioactivity: what it is, how it is produced, why it is a health hazard, and how dangerous it is. The hazard of radioactive emissions arises primarily from their energetic nature. Unlike the more mundane chemical energy associated with the burning of oil or coal, nuclear power exploits the very energies that bind the atomic nucleus. The splitting of atoms, which takes place in the heart of a nuclear reactor, also causes unstable energy balances in the fragments of the split atoms. Radioactivity is the process by which energetic stability is restored. Radioactive emissions, or radiation, are ejected from atomic nuclei, carrying away this excess energy. These emissions are able to penetrate matter, and it is this property that makes them biologically unsafe. Unlike many chemical toxins that can be neutralized, the hazard of radioactivity only disappears through natural decay, which may take hundreds, thousands, even millions, of years.

The biological hazards posed by radioactivity depend on the nature and extent of exposure to its emissions. Exposure may occur from medical radiation, routine or accidental releases of radioactivity from nuclear facilities, nuclear-weapons-testing fallout, or handling of radioactive materials for research and industrial purposes. That hazards do exist is not open to question, but their magnitude and the biological effects—especially at low exposure levels—are the subject of fierce controversy. Studies of radiation-exposed animal populations are numerous and have provided much useful information in this regard. However, for obvious reasons, no rigidly controlled studies of radiation-exposed human populations have ever been conducted. Instead, assessments of biological effects to humans have depended upon data from incidents of exposure arising out of intent, accident, or ignorance—for example, survivors of the atomic bombings of Hiroshima and Nagasaki or radium watch dial painters who unknowingly ingested lethal quantities of radium. In many cases, no quantitative data on actual exposure levels exist. Even today, radiation monitoring is an imprecise science; different types of radiation, causing varying effects to various parts of the body, are registered differently by different detection devices. Furthermore, such factors as radiation pathways and mobility and mechanisms whereby radioactivity might concentrate in organisms are not completely understood. Small wonder, then, that reputable estimates of the hazard posed by exposure to low levels of a particular type of radiation may differ by very large factors. These are the basic disagreements that, in part, fuel much of the debate over the radioactive waste problem. This chapter attempts to provide the basic information necessary to understanding that debate. It discusses the nature of radioactivity, the hazards of radiation exposure, movement of radioactivity through the environment, and the problem of determining what, if any, constitutes a "safe" level of radiation exposure.

THE NATURE OF RADIOACTIVITY

Radioactive Decay

All of the matter that makes up the universe is composed of atoms. Not all atoms are alike, however; they differ by virtue of the number of neutrons and protons that make up their nuclei.[1] An atom of hydrogen, for example, contains a single proton, while an atom of iron may contain twenty-six protons and thirty neutrons. All atoms of the same element contain an identical number of protons; they do not all contain the same number of neutrons. Atoms of the same element with different numbers of neutrons are called "isotopes." Uranium, for example, has two common isotopes—uranium–235 and uranium–238. Both isotopes have 92 protons, but the former has 143 neutrons, while the latter has 146.

There are ninety-two naturally occurring elements and several hundred naturally occurring isotopes. However, there also exist artificial elements and isotopes, created in nuclear research reactors and particle accelerators. (They are artificial only in the sense that they are manmade.) Fourteen artificial elements and many hundreds

---

1. Unfamiliar terms may be found in the Glossary.

of artificial isotopes have been discovered in this way. Most of the naturally occurring isotopes are stable—that is, they have no tendency to break up into smaller atoms. But if a large number of neutrons is added to (or removed from) the nucleus of a stable atom, the energy balance between the nuclear protons and neutrons will become uneven, and particles may be expelled from the nucleus in order to restabilize the energy balance. For example, an atom of plutonium-239, an artificial element created by adding a neutron to uranium-238, may expel an alpha particle, which consists of two protons and two neutrons. The nucleus will have two fewer protons and two fewer neutrons and will thus be an atom of uranium-235, a naturally occurring uranium isotope.[2] A lighter unstable atom, such as cesium-137, may emit a beta particle, which is an electron, and thus become an atom of barium-137m. This happens because a neutron can spontaneously convert to a proton, an electron, and a massless neutrino. The electron and neutrino will leave the nucleus, and the proton will remain behind. Thus, a new element results. This phenomenon of particle emission is called "radioactivity." The new atoms produced may be stable or unstable.[3] For the most part, artificial isotopes are unstable and decay radioactively; many unstable natural isotopes also exist, such as those of uranium. Besides alpha and beta particles, radioactive decay can also produce electromagnetic radiations similar to light but of much greater energy; these radiations are called "gamma rays" and "x-rays." Free neutrons are a fifth type of radiation. They do not occur as a result of natural radioactivity but are produced in great numbers by the nuclear reactions involved in the operation of a nuclear reactor.

Properties of Radiation

Radioactivity poses a special hazard to living things because of its peculiar nature. Radioactive emissions, or radiation, are able to penetrate matter and interact with its chemical structure. In living tissue, radioactive emissions can strip electrons from their orbits around atomic nuclei. As a result, these atoms become charged, or "ionized," and may combine with other atoms or molecules to form

2. The perceptive reader may notice that adding a neutron to uranium-238 creates plutonium-239, but eliminating two neutrons and two protons from plutonium-239 creates uranium-235. Actually, uranium-238 becomes uranium-239, which then decays by beta emission to become neptunium-239, which then decays by beta emission to become plutonium-239.

3. For example, barium-137m decays to stable barium-137 by gamma emission. This is called an "internal transition" decay because the nucleus shifts from a more energetic to a less energetic state. The excess energy is carried away as a gamma ray.

abnormal chemical complexes. Or, x-rays may be produced that can also cause atomic ionization. In some instances, the passage of a single particle may kill many cells. Radiation also disrupts molecular bonds, causing decomposition of or damage to complex molecules such as DNA. These interactions can cause a host of effects—damage to or death of exposed cells, cell mutation, induction of cancer, and injury to or even death of the exposed organism.

The ability of radioactive emissions to penetrate matter depends upon their energy and structure. As radiation passes through matter, it collides with electrons and nuclei, thereby losing energy. A heavy radioactive particle, such as an alpha particle, loses energy much more rapidly than a gamma ray. Thus, the alpha particle is much less penetrating than the gamma ray, and, in general, particles are less penetrating than electromagnetic radiation.

Because some radioactive emissions are so penetrating, it is not necessary that the radiation source be situated inside living tissue for damage to be done. A gamma ray emitter, to pick the worst example, can cause damage at distances in air of many feet. If the source is very intense, as in the case of an atomic explosion, gamma rays can cause damage at distances of hundreds or thousands of feet. This is so because gamma radiation loses very little energy as it passes through air. Beta emitters must be closer to cause damage and are relatively innocuous at distances of a few feet because electrons, being charged, are easily slowed by a few feet of air or a thin sheet of metal. Alpha emitters must be virtually in contact with tissue in order to cause injury.

All forms of radioactivity can be damaging if ingested, inhaled, or otherwise incorporated into body tissues. Alpha particles are especially dangerous in this respect. Because an alpha particle slows down so rapidly in tissue, it deposits a much greater amount of energy in a given volume of tissue than other types of radiation. Even though gamma rays are much more penetrating and irradiate greater tissue volume, the effect on individual cells is not as great as with the case of alpha particles. Beta particles fall between these two extremes in terms of effects.

Neutrons, produced in great quantities in nuclear reactors but not by radioactive materials, are themselves unstable particles. They can cause deleterious effects through nuclear reactions they may induce, transmuting stable atoms into unstable species and creating sprays of gamma rays. Furthermore, the high energy neutrons produced in nuclear reactors are extremely penetrating: an eight inch thick steel wall that would attenuate a large fraction of the radiation from a high energy gamma source would have little effect upon neutrons of equivalent energy (Nelson and Wodrich, 1974: 5).

Nuclear Fission

Nuclear fission is another kind of radioactivity: the splitting of special nuclei—certain isotopes of uranium and plutonium—into two parts of nearly equal mass. Unlike the radioactive decay discussed above, nuclear fission rarely occurs spontaneously. It can, however, be induced by neutrons. When atoms of certain natural and man-made elements, such as uraniums-233 and 235 or plutonium-239, are struck by neutrons, they may split, or fission, into two smaller nuclei, or "fission products," with the simultaneous release of several neutrons. The fission products are themselves radioactive. The decay of these fission products produces heat, but the fission process, induced and kept under control, is the dominant source of the energy produced in nuclear reactors. It takes about twenty-five billion fissions to produce the energy equivalent to the combustion of one gallon of oil.

If a sufficient number of fissionable atoms are present in a system, the neutrons released by the first fission will cause other fissions, producing more heat-releasing fission products and still more free neutrons. This process can increase geometrically and is called a "chain reaction." If kept under control, as in a reactor, it will continuously produce energy; if uncontrolled, an atomic explosion will result.[4]

The fissionable, or "fissile," atom in the current generation of light water power reactors is uranium-235. Alternatively, a nonfissile, or "fertile," isotope, such as uranium-238 or thorium-232, may capture a neutron and become transmuted into a new, perhaps fissile, material such as plutonium-239 or uranium-233. It may continue capturing neutrons, becoming an even heavier atom. All of the elements heavier than uranium—"transuranic elements"—have been created in this way, and all are unstable, decaying into lighter radioactive atoms—some less massive than uranium—by successive alpha emission down a "decay chain," finally to become the stable element lead. Table 1–1 shows such a decay chain. Transuranic elements are produced in great quantities in nuclear reactors. Some remain radioactive for many tens of thousands of years.

Different radioactive elements, or "radionuclides," decay at different rates. A measure of this decay rate is the "half-life"—that is, that period during which half of the radioactive atoms present in a sample will decay. This does not mean that after two half-lives the

---

4. It should be stressed, however, that such an explosion cannot happen in the type of nuclear power plants currently in operation.

Table 1-1.   The Uranium-238 to Lead-206 Decay Chain.[a]

Radionuclide	Radiation Emitted	Radionuclide [b] Half-life
Uranium-238		4,510,000 yr
↓ →	alpha, gamma	
Thorium-234		24.1 d
↓ →	beta, gamma	
Protactinium-234		1.2 m
↓ →	beta, gamma	
Uranium-234		247,000 yr
↓ →	alpha, gamma	
Thorium-230		80,000 yr
↓ →	alpha, gamma	
Radium-226		1,622 yr
↓ →	alpha, gamma	
Radon-222		3.8 d
↓ →	alpha	
Polonium-218 [c]		3.0 m
↓ →	alpha, beta	
Lead-214		26.8 m
↓ →	beta, gamma	
Bismuth-214 [d]		19.7 m
↓ →	alpha, beta, gamma	
Polonium-214		0.00016 s
↓ →	alpha	
Lead-210		22 yr
↓ →	beta, gamma	
Bismuth-210 [e]		5.02 d
↓ →	alpha, beta	
Polonium-210		138.3 d
↓ →	alpha, gamma	
Lead-206		Stable
none		

[a] There are three other decay chains: Uranium-235 to Lead-207 (Actinium decay series); Plutonium-241 to Bismuth-209 (Neptunium decay series); and Thorium-232 to Lead-208 (Thorium decay series).

[b] yr = year; d = day; m = minute; s = second.

[c] A small fraction of Po-218 decays to Astatine-216, which then decays to Bi-214.

[d] A small fraction of Bi-214 decays to Thallium-210, which then decays to Po-214.

[e] A small fraction of Bi-210 decays to Thallium-206, which then decays to Lead-206.

Source: HEW (1970:112).

radioactivity has completely disappeared. After two half-lives, half of the radioactive atoms present at the beginning of the second half-life will have decayed, leaving one-fourth of the original sample. After three half-lives, one-eighth of the original sample remains, and so on, ad infinitum. The radioactivity of a sample decays at what is called an "exponential rate." Figure 1–1 shows the exponential decay curve of cesium–137, a radioisotope whose half-life is thirty years. After ten half-lives, the radioactivity remaining is only about one-thousandth of the original activity, but the sample is still radioactive and may still be quite dangerous.

Radionuclides can have very different half-lives, some long, some short. One with a short half-life is highly radioactive and may transmute into another species—radioactive or nonradioactive—in periods as short as a small fraction of a second. A mass of a radionuclide with a long half-life will continue to produce radioactivity in some cases for billions of years. An arbitrary dividing line between those species considered short lived and those considered long lived is usually set at a half-life of about one hundred years.

In general, fission products are short lived, with most having half-lives of a few tens of years or less, although some, such as iodine–129, have half-lives of a million years or more. The transuranic elements produced in nuclear reactors are nearly all long lived, with half-lives greater than several thousand years. Again, however, there are exceptions; some transuranic isotopes have very short half-lives. These do not pose the great problems of waste disposal associated with the long-lived radionuclides. Appendix A contains a listing of some radioisotopes present in reactor fuel, their half-lives, and their specific activities.

Measuring Radioactivity

The unit of radioactivity is the curie, a direct measure of the radioactive disintegration rate of a particular sample of material. One curie is equal to thirty-seven billion radioactive disintegrations per second, the equivalent of the decay rate of 1 gram of pure radium.[5] Amounts of radioactivity much less than a curie can be biologically important, hence the microcurie—one-millionth of a curie—and the nanocurie—one-billionth of a curie. Amounts of radioactivity produced in nuclear reactors can be very much larger than one curie. Thus the megacurie—one million curies. A single large reactor may contain over 10,000 megacuries of radioactivity.

5. The curie is, of course, so called in honor of Madame Curie, who first extracted pure radium from uranium ore.

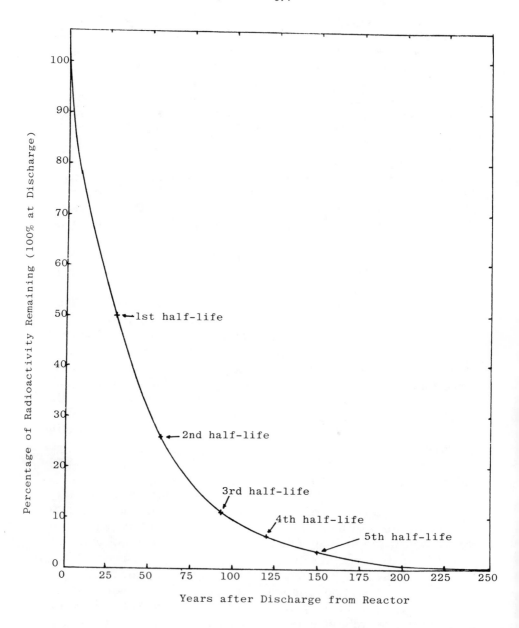

Figure 1–1. Exponential Decay Curve for Cesium–137, a Radionuclide with a Thirty Year Half-life. Note that during each half-life, the amount of radioactivity remaining declines by half.

RADIATION EXPOSURE

The effects of radiation upon living things have been extensively studied, and a large body of technical literature exists that documents a wide range of harmful consequences (see Section 2 of Bibliography). Among these are cancer, reproductive failure, genetic defects, birth abnormalities, and cell death. The precise effects of radiation exposure depend upon a number of factors—dose, dose rate, type of radiation (alpha, beta, or gamma), mode of exposure (external versus internal, inhalation versus ingestion), and age and health of the individual, among others. Many questions remain to be resolved about the significance of some of these factors, such as the exact effects of long-term exposure to low doses. Many early studies indicated that these effects were either nonexistent or minimal. Some recent evidence suggests, but does not yet conclusively prove, that low doses may cause more cancers than previously thought. The matter has yet to be resolved. (The background of the controversy is discussed in Morgan, 1978.) Nevertheless, the general life-threatening and life-damaging consequences of radiation exposure are widely agreed upon in the scientific and medical communities. Current understanding of the consequences of radiation exposure is summarized briefly in the following section.

Measuring Radiation Exposure

Exposure to radiation is generally expressed in terms of two quantities—rads and rems. A rad is a measure of radiation transfer. It is the quantity of energy carried by a radioactive emission that is absorbed in an irradiated material. (Specifically, it is that amount of radiation that deposits 100 ergs[6] of energy in one gram of material.) A rem—which stands for "roentgen equivalent man"—is a unit of radiation exposure based on the rad that incorporates a measure of biological consequences. A rem is equal to the radiation dosage in rads multiplied by a factor called the "relative biological effectiveness" of the ionizing radiation. For gamma rays and beta particles this factor is one; for alpha particles it is approximately ten. Thus, an alpha particle results in ten times as much exposure as a gamma ray of equivalent energy. By current standards, an exposure of one rem to one person is considered significant. Total radiation exposure is

6. An erg is that amount of energy expended when the force necessary to give an acceleration of one centimeter per second per second to one gram of mass (one dyne) acts through a distance of one centimeter. A 1 watt flashlight bulb (at 1 percent efficiency) produces about 100,000 ergs of energy per second.

generally expressed in thousandths of a rem, or "millirems." The rate of exposure is normally expressed as "millirems per hour, day, or year" or "rems per lifetime."

Living things have always been exposed to radiation because the environment has never been completely free of radiation. "Background radiation" from cosmic rays, uranium, radium, and thorium in common rock, and radioactive carbon and potassium in living things exists everywhere. The background level may vary in intensity by a factor of two or more depending upon locale and altitude. For example, the background level in Denver, as a result of local ores and higher altitude, is about twice that on the east coast of the United States. Typical background exposures to individuals vary from 80 to 190 millirem per year in the United States. Table 1–2 shows some sources of natural and manmade radiation exposure.

Biological Consequences of Radiation Exposure

Radiation effects to living things can be short term or long term. Short-term effects, which can result from exposure to sufficiently intense radiation over a brief period of time, are characterized as "radiation sickness." Long-term effects may take years or decades to become apparent and can result from abrupt, large radiation exposure or extended exposure to low radiation levels.

Table 1–2. Sources of Radiation Exposure in the United States, 1978
(U.S. general population exposure estimates).

Source	Millirem per Person per Year (average)	
Natural background (cosmic rays and radioactive rock)	High:	188
	Low:	77
Medical		80
Technologically enhanced (phosphate mining and milling, coal burning, rock construction materials)		5
Nuclear weapons testing and fallout	High:	8
	Low:	5
Nuclear energy		0.3
Consumer products (smoke detectors, etc.)		0.03
Total	High:	281.33
	Low:	167.33

Source: HEW (1979:15); Shapiro (1972:278).

The relationship between radiation exposure and consequent biological damage is an important matter. Many experts believe that there is no lower limit to the exposure that can cause damage—in short, that there is no "safe dose." This is undoubtedly true in the case of genetic effects; whether it is true in the case of cancer induction is a matter of great and continuing controversy.

Short-term Effects. Radiation sickness is the inevitable result of an acute dose of radiation of 100 rems or more. An "acute" dose is generally considered to be one that occurs over a period of about twenty-four hours. Any dose above 1,000 rems is invariably fatal within several hours to two weeks after exposure. Between 600 and 1,000 rems, mortality ranges from 90 to 100 percent. The mortality rate at 400 rems is about 50 percent, while below 200 rems, survival is almost a certainty (Glasstone and Dolan, 1977: 580).

The time to onset of the symptoms of radiation sickness is dose-dependent, beginning more rapidly with the greater dose. The sickness can be divided into three phases—initial, latent, and final. Initial symptoms include nausea, vomiting, intense headache, dizziness, and a general feeling of illness. Acute radiation exposure in the range of 1,000 to 3,000 rems leads to the "gastrointestinal syndrome," characterized by death within two weeks as a result of circulatory collapse. Exposure to more than 3,000 rems results in the "central nervous system syndrome," characterized by convulsions, tremor, loss of muscular control, and death within one to forty-eight hours due to respiratory failure and brain edema (DOD, 1975: 75–77). If the dose is not a lethal one, the latent period follows, with few, if any, observable symptoms. The final phase may require hospitalization of people with high exposures. Additional symptoms include skin hemorrhages, diarrhea, loss of hair, seizure, prostration, and loss of antibody production, with consequent susceptibility to infection. The final phase leads either to recovery or to death. Recovery may not be complete, however, and continued poor health is one consequence of acute exposure.

Long-term Effects. Long-term or "chronic" exposure results from extended exposure to low level radiation sources, leading to delayed effects. These effects may develop only after a latent period, which can be as long as several decades, and may include cancers, reproductive failure, genetic defects, birth abnormalities, and cell death. In particular, exposure to mobile radionuclides can give rise to these long-term effects. Experiments on beagle dogs have shown

Table 1-3.   Biological Effects of Ionizing Radiation.

Exposure Range	Chronic Exposure[a]	Acute Exposure[b]
Less than 1 rem	No observable effects; equivalent to exposure from background radiation for 5-10 years. Cancer risk: $1-2 \times 10^{-4}$ per rem for adult; greater than 4×10^{-4} per rem for fetus (may be as high as 6×10^{-4} to 6×10^{-3} per rem for children; 7×10^{-3} for adults).	No observable short-term effects.
1-50 rems	Chromosomal aberrations in blood. 0.3-30 leukemia cases per 10,000 person-rem observed in this exposure range; 0.5-1.2 thyroid cancers per 10,000 person-rem observed. Occupational exposure range.	Slight blood changes; decreased head circumference and increased leukemia risk in fetus.
50-100 rems	Approximate doubling dose for spontaneous mutations.	Mild symptoms of radiation sickness possible. Vomiting in 5 percent of those exposed to 100 rems within three hours.
100-200 rems	Approximate doubling dose for cancer.	Vomiting in 5 percent of those exposed to 100 rems to 50 percent of those exposed to 200 rems within three hours. Also, fatigue, loss of appetite, moderate blood changes that persist. Recovery within several weeks. Increased cancer risk; cataracts possible.
200-600 rems	Limited experience with regard to chronic exposure over 200 rems. Large increase in incidence of leukemia and other cancers. Uranium miners exposed to 700-1,000 rads, with maximum exposures estimated to be as great as 10,000 rads. Excess of lung cancer deaths may reach 600-1,100 in a population of 6,000 miners.	Vomiting: 50 percent at 200 rems within three hours; 100 percent above 300 rems within two hours. Also, loss of hair, other symptoms of radiation sickness. Death in 0-80 percent of those exposed within two months from hemorrhage and infection; recovery for survivors in one to twelve months.

600–1,000 rems	Vomiting within one hour, severe blood changes, hemorrhage, infection, loss of hair, damage to bone marrow. Death in 80–100 percent of those exposed within two months from hemorrhage and infection. Long convalescence for survivors.
1,000–3,000 rems	Vomiting within thirty minutes; radiation sickness. Gastrointestinal syndrome within five to fourteen days, including diarrhea, fever, severe blood changes, damage to bone marrow. Death in 90–100 percent of those exposed within two weeks due to circulatory collapse.
More than 3,000 rems	Vomiting within thirty minutes. Central nervous system syndrome within two days, including convulsions, tremor, loss of muscular control. Death in 100 percent of those exposed within one to forty-eight hours due to respiratory failure and brain edema.

[a]Exposure over extended time period.
[b]Exposure over period of twenty-four hours or less.

Sources: Morgan (1978); Rotblat (1978); Schurgin and Hollocher (1975); Glasstone and Dolan (1977:580); DOD (1975: 75–77); Shapiro (1972).

that exceedingly small quantities of plutonium–239 (three-millionths of a gram or about 200 nanocuries) embedded in lung tissue can induce lung cancer (Bair and Thompson, 1974: 720). Moreover, some internal organs are more sensitive to radioactivity than others, and some radionuclides are "organ seeking"—that is, due to their chemical nature they settle preferentially in certain organs. For example, strontium–90, chemically similar to calcium, settles in bone. Radioactive iodines, such as iodine–131, collect in the thyroid gland. Each of these radionuclides is capable of inducing cancer after a period of latency.

Radiation may also cause serious damage to reproductive cells by causing chromosomal alteration or breakage, resulting in unfavorable genetic mutations that may appear only in future generations. According to a Ford Foundation study (Ford–MITRE, 1977), if the present legal limit for manmade, nonmedical radiation were to be reached through the use of nuclear power, an increase of 0.2 to 5.5 percent in the present incidence of ten genetic abnormalities per one thousand live births could result in the first generation alone. Damage would continue out to twenty or thirty generations, although the bulk would be inflicted on the first three or four following irradiation. If this level of radiation exposure were maintained during subsequent generations, genetic disease might occur at a rate 2 to 50 percent higher than that presently observed (Ford–MITRE, 1977: 172).

Radiation can also cause serious in-utero effects by damaging the chromosomes of fetal cells. Such damage may cause severe morphological or mental defects, especially if exposure occurs during the first week after conception, when the total number of fetal cells is quite small. A small amount of damage early in pregnancy may have large delayed and cumulative effects. Table 1–3 shows the health effects of exposure to various levels of radiation.

PATHWAYS IN THE ENVIRONMENT

Radioactive atoms are chemically identical to their nonradioactive counterparts. Thus, when releases of radioactivity into the biosphere take place, the radionuclides may become integrated into various natural chemical and biological pathways and can, as a result, become concentrated in plants and animals. Many of these "concentration mechanisms" have been extensively studied, and a considerable body of empirical knowledge exists regarding chemical pathways into the food chain (see Section 2 of Bibliography). Nonetheless, accidental and unsuspected concentrations may occur.

Radioactivity can enter the environment in many ways. There are intentional discharges of radioactive effluents from various nuclear facilities. Dilution and dispersion of these effluents are often the principal mechanisms relied upon to reduce concentrations of radioactivity in air and water to federally allowed limits. There may also be accidental releases, such as occurred during the Three Mile Island nuclear power plant accident in March 1979. Even if concentrations of radionuclides are below the prescribed limits, the materials can be reconcentrated in the food chain. For example, strontium-90, if present in a cow's food or drinking water, may be ingested and concentrated in the cow's milk. The milk can constitute a hazard if consumed by humans. Because the strontium is chemically similar to calcium, it will seek out and become incorporated in bone, where its radiations may induce leukemia.

Marine organisms that feed on ocean bottom sediments and such organisms as plankton and algae are able to concentrate radionuclides to levels greatly exceeding those in the water they live in. The concentration factor is a measure of this increase. For example, the mean concentration factor for fallout plutonium in the floating alga *Sargassum* is 20,000 (Bowen, 1974: 48). The concentration factor for strontium-90 in clam shells has been found to be as much as 65,000 (Price, 1971: 9). Other organisms can also concentrate radionuclides to high levels. A study of mallard ducks on the Atomic Energy Commission's (now Department of Energy) Hanford Reservation found cesium-137 concentrations in the ducks' flesh to be 2,000 to 2,500 times that in their food (Price, 1971: 12). Another study demonstrated the potential hazard of unsuspected contamination: it found elevated levels of radioactivity in coastal marine life located over 300 miles from Hanford, in Willipa Bay at the mouth of the Columbia River (Fix, et al., 1977: 16).

RADIATION EXPOSURE STANDARDS
AND GUIDELINES

Exposure Limits

Quantifying radiation exposure effects and setting allowable exposure standards and radiation release guidelines have been and continue to be complicated procedures subject to much controversy among experts. The first radiation exposure standards were established in the 1930s, only after it became apparent that unlimited exposure to radiation was unsafe. As more information accumulated, allowable exposure levels gradually were reduced as standards became stricter. Permissible dose levels—measured in rads or rems per week or year—were based upon the assumption—called the "threshold hypothesis"—that below a certain dose level, the risks would be small and could probably not be identified statistically or by other means. The most recent exposure standard, established in the late 1950s, was based on the observation that, because the average annual exposure per person in the United States was 120 millirem due to natural background radiation plus 50 millirem resulting from medical radiation, the effects from an additional 170 millirem per year would most probably be undetectable. This remains the exposure standard for the general public under normal conditions. However, radiation workers are allowed up to five rems per year over a limited number of years, and in the event of an accidental radiation release, the allowable radiation dose to the public has been set by the Nuclear Regulatory Commission at one-half rem. Table 1–4 shows the dose limit recommendations of the National Council on Radiation Protection and Measurements, an advisory board on radiation exposure.

Low Level Radiation Exposure

Evidence suggests that the radiation threshold hypothesis is incorrect, and in recent years the threshold hypothesis has been increasingly supplanted by the "linear hypothesis." Here one assumes that effects from low level radiation exposures can be linearly extrapolated from effects at high exposures. Using this hypothesis as a conservative upper bound, one of the more definitive studies on radiation effects conducted by the National Academy of Sciences concluded that the exposure of the entire population of the United States to the 170 millirem per year limit over a thirty year period could cause from 3,000 to 15,000 excess cancer deaths annually over the current domestic cancer death rate, with 6,000 deaths per year being the most likely figure, plus an increase of 5 percent in the ill-health of the population (BEIR Report, 1972: 2). The current domestic cancer death toll, for comparison, is about 360,000 per year.

Recent information on the effects of low level ionizing radiation indicates that in some circumstances even the linear hypothesis may be invalid. It has been suggested that the hazard of low, extended doses of radiation may actually be greater than that of short, acute exposures (fifty rems or more) because cancer-prone cells that might be killed by acute doses could survive low exposures and eventually become cancerous (Morgan, 1978: 35). Laboratory studies of this effect have been inconclusive, for in some instances, animals exposed to low, extended doses of radiation have recovered with no long-term

Table 1—4.   Radiation Dose Limits Recommended by the National Council on Radiation Protection and Measurements, 1971 *(Adopted by the Nuclear Regulatory Commission).*

Occupational Exposure Limits:	
Whole body (prospective)	5 rems in any one year
Whole body (retrospective)	10–15 rems in any one year
Whole body (accumulation to age N years)	$(N - 18) \times 5$ rems
Skin	15 rems in any one year
Hands	75 rems in any one year (25/quarter)
Forearms	30 rems in any one year (10/quarter)
Other organs, tissues, and organ systems	15 rems in any one year (5/quarter)
Fertile women (with respect to fetus)	0.5 rem in gestation period
Dose Limits to the Public or Occasionally Exposed Individuals:	
Individual or occasional	0.5 rems in any one year
Students	0.1 rem in any one year
Population Dose Limits:	
Genetic	0.17 rem average per year
Somatic	0.17 rem average per year
Emergency Dose Limits—Lifesaving:	
Individual (older than forty-five years if possible)	100 rems
Hands and forearms	200 rems additional (300 rems total)
Emergency Dose Limit—Less Urgent:	
Individual	25 rems
Hands and forearms	100 rems total
Family of Radioactive Patients:	
Individual (under forty-five years)	0.5 rem in any one year
Individual (over forty-five years)	5 rems in any one year

Source: NCRP (1971: table 6).

effects. However, studies in progress on radiation workers at government defense facilities have cast doubt on the linear hypothesis, instead pointing to a higher dose-risk relationship at low exposure levels than would be assumed from the linear hypothesis (*Nucleonics Week*, 1979a: 2; *New Scientist*, 1979: 75). These studies also indicate that no threshold whatever exists for radiation effects—that is, all radiation doses are harmful. As a result, there is a growing body of responsible opinion that allowable exposure levels to radiation workers and the general public should be lowered significantly.[7] Figure 1−2 shows possible dose response curves.

Limits on Radioactivity in the Environment

The amount of radioactivity that is allowed to enter the environment is, for the most part, strictly controlled. Guidelines, established by the Nuclear Regulatory Commission and the Environmental Protection Agency, set maximum allowed limits for the intentional release of radioactive materials into air and water. The guidelines have been established in the following manner:

1. A maximum annual safe radiation exposure for specific organs and the whole body is determined as a result of experiment and empirical observation.
2. Models are then constructed that determine exposure routes and rates for radioactive materials in these organs.
3. These models allow a correlation to be drawn between permissible exposure and the quantity of inhaled or ingested radioactivity that would cause this exposure, which in turn sets a maximum on the quantity of radioactivity allowed in air and water intended for unrestricted public use. This quantity is known as the "maximum permissible concentration," or MPC, for radionuclides in air and water. The MPCs for several radionuclides may be found in Table 1−5.

Legally, no nuclear facility is allowed to discharge radioactive effluents into the environment in concentrations that would result in these values being exceeded, and actual releases are low enough under the majority of circumstances to preclude even accidentally exceeding the MPC. These maximum limits should not be viewed as

7. The size of the decrease finally recommended by the Nuclear Regulatory Commission—if indeed, a reduction is recommended at all—may not necessarily wholly reflect concern for health and safety as much as for economic factors, since according to the Atomic Industrial Forum, the lobbying association for the nuclear industry, a tenfold reduction in exposure standards could cost the industry as much as $500 million per year (*Nucleonics Week* 1979a: 2).

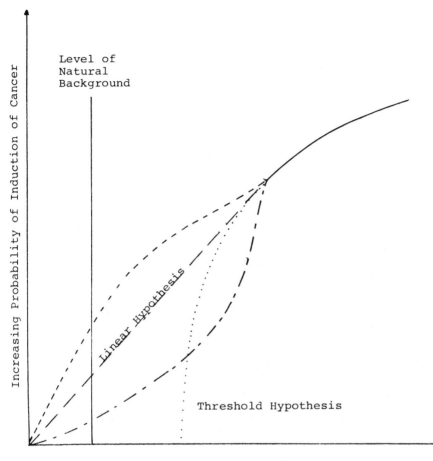

Figure 1–2. Possible Dose Response Curves for Exposure to Low Level Radiation are Depicted on this Graph. The solid line indicates the high dose region in which the response to a given radiation dose has been empirically drawn from observation. The threshold hypothesis assumes that no response occurs below a certain threshold dose. The linear hypothesis assumes a linear (straight line) dose response down to zero dose. The upper curve assumes a greater than linear response; the lower curve, a less than linear response. The critical factor is the slope of the curve in the natural background region.

Source: NAS (1979).

Table 1—5. Maximum Permissible Concentrations for Several Radionuclides Found in Spent Reactor Fuel.

Radionuclides	Maximum Permissible Concentration[a] (curies per cubic meter)[b]	
	Air	Water
Krypton–85	3×10^{-7} [c]	—[d]
Strontium–89	3×10^{-10}	3×10^{-6}
Strontium–90	3×10^{-11}	3×10^{-7}
Iodine–129	2×10^{-11}	6×10^{-8}
Iodine–131	1×10^{-10}	3×10^{-7}
Cesium–135	2×10^{-8}	1×10^{-4}
Cesium–137	2×10^{-9}	2×10^{-5}
Radium–226	3×10^{-12}	3×10^{-8}
Uranium–235	2×10^{-11}	3×10^{-5}
Uranium–238	3×10^{-12}	4×10^{-5}
Plutonium–238	7×10^{-14}	5×10^{-6}
Plutonium–239	6×10^{-14}	5×10^{-6}
Americium–241	2×10^{-13}	4×10^{-6}
Americium–243	2×10^{-13}	4×10^{-6}

[a] For soluble material; MPCs for insoluble materials are generally higher.

[b] An adult drinks about 0.8 cubic meters of water per year and inhales about 7,300 cubic meters of air each year.

[c] The notation is read as powers of 10. For example 3×10^{-7} is the same as 0.0000003, or 3 ten-millionths.

[d] Krypton–85 is an inert gas; therefore, it does not remain in water.

Source: U.S. Code of Federal Regulations, Title 10, Part 20, updated through October 1975.

definitive, however. The values are conservative but are nonetheless based upon controversial data and imperfect models. That is not to say that the limits are too high or too low, but rather that given the available evidence, they are probably not unreasonable.

Allowed limits for radionuclides in foodstuffs, set by the Environmental Protection Agency, although strictly established, may be difficult to control. For example, although there is a limit to the maximum amount of strontium–90 allowed in milk for uncontrolled human consumption, the hazard varies among individuals according to age, health, and other factors. In addition, restrictions on consumption of contaminated foodstuffs are applied only in instances of known radiation releases. Under conditions of an accidental but undetected release, no testing for radioactivity in foodstuffs would ordinarily take place. Furthermore, even though intentional radiation releases may be far below allowed limits at the point of origin, concentration may occur in organisms many hundreds of miles from the source of radioactivity, possibly exposing individuals, distant and unaware, to unsafe levels of radiation. Table 1-6 gives the allowed limits for several radionuclides in milk. Table 1-7 gives dose conversion factors for exposure to some radionuclides.

This brief overview of the nature and hazards of radioactivity is intended only as background material to the radioactive waste problem. Understanding of these characteristics is crucial to understanding the waste problem, however, for they lie at the core of the problem. It is the ability of radioactive emissions to penetrate living tissue that allows not only the beneficial uses of radiation, such as x-rays and radiation therapy, but also, under certain circumstances, to potential health hazards, such as cancer, genetic abnormalities, and birth defects.

The specific level at which exposure to small amounts of radiation becomes dangerous is the subject of widespread debate and disagreement, but there exists a general consensus that accidental exposure to radiation, even in small doses, may be dangerous. Ordinarily, human beings are exposed only to natural background radiation from cosmic rays, uranium, radium, and thorium in rock, and radioactive carbon and potassium in living things, and to occasional medical radiation. In recent decades, these sources of radiation have been slightly enhanced by radioactive fallout from the testing of nuclear weapons. The large-scale fissioning of uranium in nuclear reactors, however, carries with it the potential for a sizable increase in the quantity of

Table 1-6. Maximum Permissible Concentration in Milk of Several Radionuclides.[a]

Nuclide	Maximum Concentration (microcuries/liter)
Strontium–89	3.7[b]
Strontium–90	0.155[b]
Iodine–131	0.084[c]
Cesium–137	2.4[d]

[a] Assumes consumption of one liter of milk per day over a period of one hundred days, with exposure limit of 10 rems mean dose to relevant organ over first year and lifetime strontium–90 dose not to exceed 15 rems. Standards established by Federal Radiation Council, 1961–1965.

[b] Dose to bone marrow.

[c] Dose to 2 gram thyroid.

[d] Dose to whole body.

Source: NAS (1973: 82).

Table 1–7.   Typical Dose Conversion Factors for Exposure to Selected
Radionuclides.

External Radiation Dose from Exposure to Contaminated Ground [a]

Nuclide	Dose (rem per curie/square meter per day)
Cobalt–60	848
Cesium–134	530
Cesium–137	186
Plutonium–238	0.90
Plutonium–239	0.38
Plutonium–240	0.78
Americium–241	20.60

Dose Equivalent per Unit of Activity Ingested (rem per microcurie) [b]

Nuclide	Bone	GI Tract	Lungs	Thyroid	Total
Co–60	0.003	0.04	0.003	0.0031	0.0044
Sr–90	1.2	0.078	—	0.0060	0.095
I–131	0.00036	0.00005	0.00029	1.8	0.0009
Cs–137	0.068	0.026	0.020	0.067	0.049
Pu–239	1.1	0.2	—	0.0037	0.096
Am–241	39	0.22	0.00013	3.3	0.13

Dose Equivalent per Unit of Activity Inhaled (rem per microcurie)

Nuclide	Bone	GI Tract	Lungs	Thyroid	Total
Co–60	0.051	0.029	1.3	0.06	0.08
Sr–90	3.0	0.014	0.0099	0.015	0.24
I–131	0.00025	0.000039	0.0024	1.1	0.00062
CS–137	0.045	0.016	0.016	0.045	0.033
Pu–239	1800	0.12	590	6.0	110
Am–241	1900	0.13	630	6.1	180

[a] Spread uniformly on the ground; photon dose only.

[b] A discussion of associated data and assumptions may be found in G.G. Killough et al., *Estimates of Internal Dose Equivalent to 22 Target Organs for Radionuclides Occurring in Routine Releases from Nuclear Fuel Cycle Facilities*, vol. 1, NUREG/CR-150, ORNL/NUREG/TM-190 Oak Ridge, TN: Oak Ridge National Laboratory, (June 1978). Volume 2 by D.E. Dunning et al., NUREG/CR-0150/ v. 2, ORNL/NUREG/TM-190/v.s. is in preparation.

Sources: NRC (1975: Appendix VI, C-6); NAS (1979: 118).

easily accessible radioactive material in storage, and, hence, in the potential for accidental releases of radioactive materials into the environment. Inevitably, such releases will lead to an increase in the radiation dose to specific populations. Indeed, as is discussed in Chapter 4, accidental releases of radioactivity are not uncommon, although, to date, deleterious effects have been minimal. For more detailed information about the hazards of radiation exposure, the reader should consult the references listed in Section 2 of the Bibliography. In the following chapter we describe how the radioactive wastes that are the subject of this book are produced.

Chapter 2

The Nuclear Fuel Cycle: How Radioactive Wastes Are Produced

Uranium must be dug out of the ground and processed in various ways before it can be used as fuel in a nuclear reactor. After it is "burned" in a reactor, the residual radioactive materials must be carefully handled and ultimately very carefully disposed of. The set of activities that begins with uranium ore and ends with radioactive wastes is called the "nuclear fuel cycle" and is the source of a large fraction of the radioactive waste currently being generated in the United States.

The fuel cycle consists of mining uranium ore; milling the ore to produce uranium oxide; enriching the uranium product; fabricating the enriched uranium into fuel elements; irradiating the fuel elements in a nuclear reactor; removing the spent fuel from the reactor followed by reprocessing or storage; disposing of the reprocessed wastes or spent fuel; and finally, decontaminating and decommissioning the reactor after the end of its useful operating life. These steps are shown schematically in Figures 2–1 and 2–2.

Each of these steps produces radioactive wastes, but some much more than others. Mining, milling, enriching, and fabricating uranium fuel produces low level radioactive effluents in liquid and solid form. These effluents are either discharged into the environment or into holding ponds or buried in special dumps.

Power production from the fuel and the procedures that follow produce the greatest quantities of waste. In the reactor the fuel becomes intensely radioactive and, upon removal, must be placed in a pool of water for several months to allow the highly active, short-lived radionuclides to decay away. If reprocessed, large volumes of

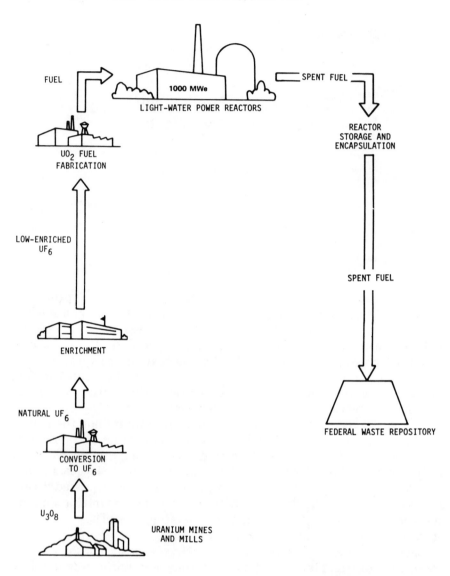

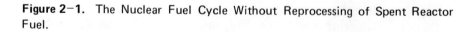

Figure 2-1. The Nuclear Fuel Cycle Without Reprocessing of Spent Reactor Fuel.

Source. NRC (1976b: 3-2).

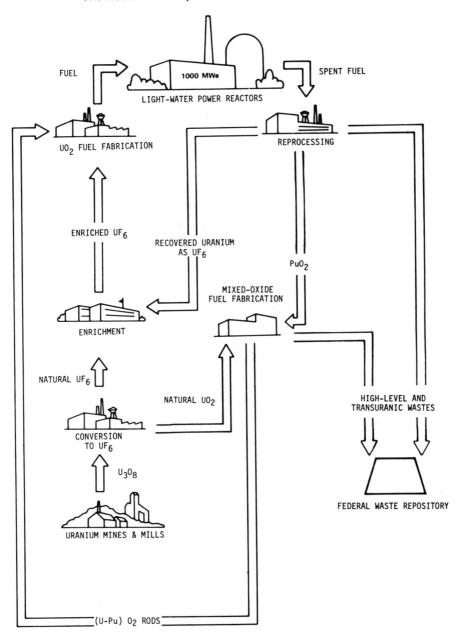

Figure 2–2. The Nuclear Fuel Cycle with Reprocessing. The uranium and plutonium extracted from spent reactor fuel are recycled into new fuel.

Source: NRC (1976b: 3–5).

Table 2-1.   Radioactive Wastes Generated Annually in the Fuel Cycle of a 1,000 Megawatt (Electric) Light Water Reactor with a 30 Metric Ton Fuel Off-load.[a]

Uranium mining and milling	96,000 metric tons of uranium mill tailings with 0.7 nanocuries of activity per gram 245,000 metric tons of tailings solutions Radon gas and daughters Low level radioactive dust and liquid effluents
Conversion to UF_6	1,200 cubic feet of low level solids, liquids, and sludge
Enrichment to 3-4 percent uranium-235	5,500 metric tons of enrichment tails containing 0.2 percent uranium-235 Low level solids: gaseous diffusion, 50 cubic feet; gaseous centrifuge, 2,900 cubic feet Low level liquids Airborne uranium (small quantities)
Fuel rod fabrication	750 cubic feet of low level solid wastes Process off-gases Low level liquids containing uranium, thorium, and protactinium
1,000 MWe Light Water Reactor	30 metric tons of spent fuel containing approximately 300 kilograms of U-235, 250 kilograms of plutonium isotopes, plus a variety of fission products; volume is 390 cubic feet Low level solids: boiling water reactor, 46,000 cubic feet with 38 millicuries of radioactivity per cubic foot; pressurized water reactor, 26,500 cubic feet with 43 millicuries of radioactivity per cubic foot Low level liquid effluents Reactor decontamination and decommissioning (only once during reactor lifetime): mothballing, 2,000 cubic feet; entombment, 80,000 cubic feet; dismantling and removal, 300,000–800,000 cubic feet
Reprocessing and mixed oxide fuel fabrication	Approximately 1.5 gallons of high level liquid waste per kilogram of spent fuel reprocessed, containing thousands of curies of radioactivity per gallon; volume is 240 cubic feet; 13.5 million curies 270 kilograms of plutonium; 3.8 million curies 530 cubic feet of spent fuel cladding hulls; 870,000 curies 200 cubic feet of low level wastes 1,060 cubic feet of transuranium-contaminated waste; 1.7 million curies

Table 2−1. continued

Spent fuel storage	6 cubic feet of low level solids and liquids
	Fission product gases
	Low level solids and liquids
High level waste repository	Transuranium-contaminated waste
	0.6 to 1 acre required for disposal of 30 metric tons of spent fuel or waste equivalent

[a]1,250 MWe reactor with 80 percent capacity factor; 1 metric ton = 1,000 kilograms = 2,200 pounds.

Source: IRG (1979:D-6); NRC (1977a; 1976b:3-15).

highly radioactive liquids will be generated; otherwise, the spent fuel itself must be disposed of as waste.

Dismantling old reactors will also produce large volumes of radioactive rubble. All of these waste products of the nuclear fuel cycle require safe handling and disposal. Table 2−1 gives the quantities of radioactive waste generated each year by one large light water reactor.

RADIOACTIVITY AND RADIOACTIVE WASTE

Radioactive wastes are classified according to the activity and nature of the contained radioactivity. Three different classifications exist— low level, high level, and transuranium-contaminated waste. The first two categories of waste differ only in degree of concentration (or dilution) of radioactivity; the nature of the radioactivity in each type of waste is generally the same. The third category was established because wastes contaminated with long-lived alpha emitters require special handling.

Low Level Wastes

Low level wastes generally average less than one curie of activity per cubic foot of material, or less than 10 nanocuries (billionths) of transuranic contamination per gram (roughly 30 microcuries per cubic foot). Low level waste is generated in many parts of the nuclear fuel cycle.[1] It typically consists of contaminated machinery, gloves, aprons, tissues, paper, and so forth and is produced in tremendous volume. The Environmental Protection Agency has estimated that by the year 2000, approximately one billion cubic feet of such

---

1. Low level wastes are also generated in great quantity by nuclear research facilities, universities, hospitals, and industry.

wastes requiring disposal will have been generated. Such a volume would, roughly, cover a four lane highway from coast to coast to a depth of one foot (House of Representatives, 1976: 4).

The amount of radioactivity in low level wastes averages less than one curie per cubic foot. The waste may, however, be contaminated to many times that level but still be defined as low level because of bulk or mixing. Low level wastes may contain items contaminated with "hot spots," where concentrations of radioactivity may be quite high. The normal treatment for these wastes has been burial at licensed facilities in shallow trenches with unsealed bottoms. As will be seen in Chapter 4, this handling has not always provided the required confinement.

High Level Wastes

High level wastes are officially defined as the waste streams that result from the reprocessing of spent reactor fuel. These wastes are radioactive liquids from the chemical solvent extraction reprocessing procedure and liquids having comparable radioactive concentrations produced in later cycles of the process. These liquids generally contain hundreds to thousands of curies per gallon and, if solidified to salt cake, much higher concentrations. Spent fuel reprocessing, however, has been indefinitely deferred by presidential decision, and therefore spent fuel has unofficially come to be considered high level waste as well. The amount of radioactivity in these wastes is great enough to generate an appreciable quantity of heat.

Transuranium-contaminated Waste

Transuranium-contaminated waste contains more than 10 nanocuries of transuranic nuclides per gram of material. Until 1970, such waste was routinely handled and buried in the same manner as low level waste. Since then, most of this material has been packaged in barrels and stored, awaiting a disposal method appropriate for long-lived waste.

GENERATION OF RADIOACTIVE WASTES

As previously discussed, wastes are produced at all points in the nuclear fuel cycle.

Mining

About 20,000 tons of ore containing 40 tons of uranium oxide are mined daily in the United States, with 87 percent of the production coming from Wyoming, New Mexico, Colorado, and Utah. The mining operations produce slightly radioactive dust and release radon gas into the atmosphere. These effluents cause low level contamination of local air and water, with dilution and dispersion being relied upon to minimize the hazards. In the past, underground mining operations posed a serious threat to the health of uranium miners as a result of radon gas buildup in poorly ventilated mines. The U.S. Public Health Service has estimated that, of 6,000 uranium miners in the United States, an excess of 600 to 1,100 have died or will die as a consequence of lung cancer induced by heavy exposure to radon gas and its daughter products during the 1940s and 1950s (Schurgin and Hollocher, 1975: 26). Present-day underground mining operations tend to be safer as a result of improved ventilation, although some critics believe that exposure levels to radon are still excessively high (Schurgin and Hollocher, 1975: 30).

Refining

The uranium ore is sent to mills, where uranium oxide is extracted from the ore. The resulting semirefined product (U_3O_8) is known as "yellowcake." Crushing, grinding, and chemical processing of the ore produces low level airborne contaminants. Among these, radon gas and radium-containing particulates may constitute a health hazard if inhaled in significant amounts. The spent ore, depleted in uranium, is discharged into settling ponds as a water solution of finely ground material, still containing about 0.7 nanocuries of radioactivity per gram (Hollocher and MacKenzie, 1975: 45). These sediments gradually dry out as the water evaporates from the ponds. Left behind is a mass of grey, fine-grained sand called "mill tailings." Even though the radioactivity level in mill tailings is quite small, the tailings continuously generate radioactive radon gas, at up to 500 times the natural background rate. Radon is part of a decay chain that begins with uranium–238 (half-life four and a half billion years) and ends with thorium–230 (half-life 80,000 years). (This decay chain was presented in Table 1–1.) Radon and its daughter products are responsible for the high incidence of lung cancer among uranium miners. Radon has a short half-life, and once inhaled, the daughter products become lodged in the lungs, where they continuously irradiate lung tissue. Even though the contribution of radon to the general level of environmental radioactivity is small, the long half-life of uranium–238 ensures that mill tailings will produce the gas virtually forever. In fact, if the tailings are not isolated from the environment, after 100,000 years they will become the greatest source of radiation exposure from the nuclear fuel cycle (Carter, 1978b: 191). These tailings, already produced in very large quantities, have not, to date, been adequately isolated. Indeed, they have been irresponsibly managed (see Chapter 4).

Enrichment

The yellowcake produced by milling is chemically processed by one of two methods to produce uranium hexafluoride gas (UF_6). The dry hydrofluor process produces waste solids containing long-lived alpha-emitting radionuclides; these wastes are disposed of in low level burial grounds. The wet method generates a liquid stream containing dissolved radioactive solids. The waste stream, called raffinate, is dumped into a settling pond, leaving a sludge as the water evaporates. The sludge contains small quantities of radium, thorium, and uranium, all long-lived radionuclides, but is handled as low level waste. Cooling water used in the gasification process also deposits small quantities of uranium into nearby streams.

The UF_6 gas must be processed to increase the uranium–235 content. Natural uranium is only 0.7 percent uranium–235. Although several enrichment processes exist, only the gaseous diffusion process has been used on a large scale in the United States. It is a complex, expensive process requiring major industrial installations. The gas is diffused through a series of permeable membranes; the U–235, being lighter than U–238, diffuses through the membranes more rapidly. The diffusion process enriches the U–235 content of the gas to the 2 to 4 percent required for reactor operation. Only small quantities of radioactive effluents are produced in the enrichment process, primarily uranium, discharged into the atmosphere, and low level liquids, released into holding ponds. New enrichment technologies currently being developed may produce increased or decreased quantities of waste. Laser enrichment generates virtually no effluents, while centrifuge enrichment may produce many times more waste than gaseous diffusion (APS, 1977: S165).

Fuel Rod Fabrication

The enriched uranium hexafluoride gas is converted into solid uranium dioxide, which is then compressed and formed into fuel pellets (Figure 2–3). The fuel pellets are loaded into zirconium alloy tubes that are sealed and assembled into fixed arrays called fuel assemblies. Three types of waste result from this procedure—process off-gases, treated before release in order to remove certain radioactive constituents; a liquid waste stream containing uranium, thorium, and protactinium, dumped into holding ponds and allowed to settle and form a sludge; and solids, incinerated and then buried at low level waste disposal sites.

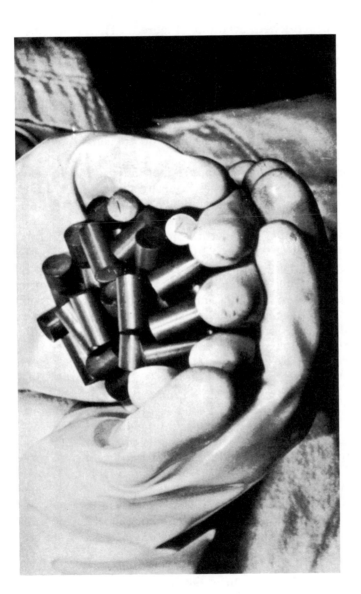

Figure 2–3. Slightly Enriched Uranium Dioxide Fuel Pellets Used in a Light Water Reactor.

Source: AEC (1973:1–5).

Reactor Operation

The operation of a nuclear reactor produces the most significant quantities of radioactive waste. A large light water reactor, the type commonly in operation in the United States, contains about 90 to 100 metric tons of enriched uranium (Figure 2-4). As the chain reaction in the reactor core proceeds, uranium-235 atoms are fissioned. The fragments, or "ashes," are intensely radioactive. Some of the nonfissile uranium-238 atoms are transmuted into transuranic elements. A few of these heavy elements, in particular plutonium-239 and 241, will fission and, as they accumulate, contribute to energy generation; the remainder, being nonfissile, build up in the reactor fuel. At the end of fuel life, about 30 percent of the produced energy comes from plutonium fission.

During reactor operation, the fission products and nonfissile heavy elements build up and ultimately reach concentrations that interfere with the efficiency of the chain reaction. When the amount of uranium-235 remaining in the fuel drops below 1 percent, the fuel elements are removed from the reactor core and replaced with fresh assemblies. One-third of the reactor fuel load—about thirty metric tons—is exchanged annually. At the time of discharge, the irradiated spent fuel from a reactor off-load contains almost 180 megacuries of radioactivity per metric ton and generates 1.5 megawatts of thermal power (heat) per metric ton. It must be cooled constantly to remove the heat. After 150 days, decay of the short-lived radionuclides has reduced the activity to four megacuries per metric ton and the thermal output to 20 kilowatts per metric ton (JPL, 1977: 4-13). Even so, the spent fuel still contains at least fifty significantly active radionuclides (NRC, 1977a: 25). The radioactivity and heat content of spent fuel as a function of time are shown in Figures 2-5 and 2-6 and Tables 2-2 and 2-3 (see also Appendix A).

The spent fuel assemblies are stored in pools of water sited adjacent to the reactor building (Figure 2-7). After irradiation, the spent fuel still contains significant quantities of uranium-235 and newly created plutonium-239, both potentially usable nuclear fuels. These elements can be extracted from the fuel by chemical reprocessing; the uranium must be returned to the enrichment plant for reenrichment. Both can then be used in the fabrication of new fuel rods. Originally it was intended that spent fuel be allowed to "cool" for five or six months until radiation levels had decayed sufficiently to allow safe handling and, only then, be reprocessed. However, for political and economic reasons, no spent fuel has been or for the foreseeable future will be reprocessed.

Figure 2–4. Fuel Loading at Baltimore Gas and Electric's Calvert Cliffs Reactor.

Source: Combustion Engineering, Inc.

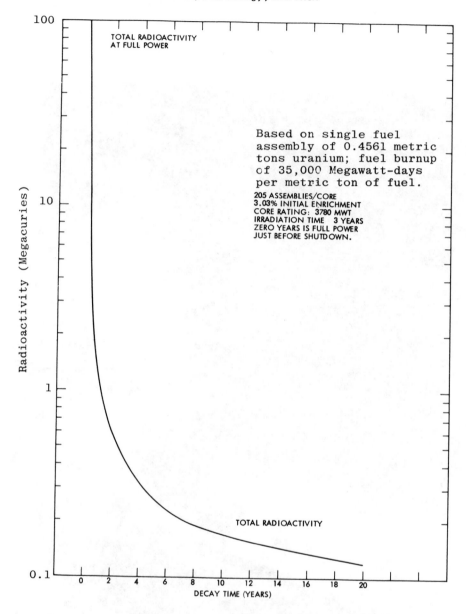

Figure 2-5. Total Radioactivity for Fission Products and Transuranics Contained in one PWR Spent Fuel Assembly as a Function of Decay Time.

Source: JPL (1977:4-3).

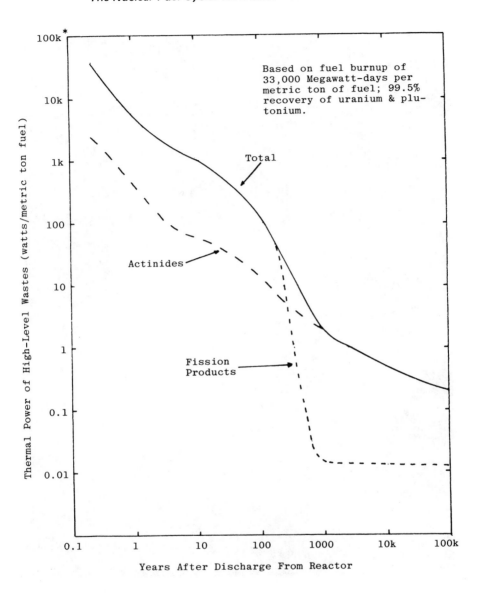

Based on fuel burnup of
33,000 Megawatt-days per
metric ton of fuel; 99.5%
recovery of uranium & plutonium.

Figure 2–6. Thermal Power Released by the Radioactivity of High Level Waste from Reprocessing 1 Metric Ton of Irradiated Light Water Reactor Fuel.

Source: Hollocher (1975:232).
Note: *k = thousand.

Figure 2–7. Spent Reactor Fuel from the High Neutron Flux Experimental Reactor. The glow is Cerenkov radiation, induced by gamma rays passing through the water in which the fuel is immersed.

Source: E.I. DuPont DeNemours and Co.

Table 2-2.   Radioactive Inventory of One Metric Ton of Spent Reactor Fuel.

	Radioactivity (in curies per metric ton)[a]						
	150 d	1 yr	5 yr	10 yr	50 yr	100 yr	500 yr
Fission Products	4,500,000	560,000	480,000	320,000	110,000	35,000	50
Actinides (including uranium isotopes)	140,000	133,000	110,000	85,600	18,500	6,960	2,940
Total radioactivity	4,640,000	693,000	590,000	405,600	128,500	41,960	2,990

Table 2-2. continued

	1,000 yr	10,000 yr	100,000 yr	1,000,000 yr	10,000,000 yr
Fission Products	22	20	16	3.6	0.08
Actinides (including uranium isotopes)	1,730	450	40	17.0	b
Total radioactivity	1,752	470	56	20.6	

[a]Fuel burnup of 33,000 MWd/metric ton. Numbers are rounded. d = day; yr = year.
[b]Daughter products of long-lived actinides reach secular equilibrium at approximately 15 curies.
Source: Hollocher (1975:226); JPL (1977:A-8); Bell (1973:table A-IV-1).

Table 2–3. Thermal Output of One Metric Ton of Spent Reactor Fuel.

	Thermal Output (in watts per metric ton)[a]					
	150 d	1 yr	10 yr	50 yr	100 yr	500 yr
Fission Products	20,800	12,100	1,060	353	106	0.07
Actinides	670	409	208	217	193	96
Total	21,470	12,509	1,268	570	299	96.07

Table 2–3. continued

	1,000 yr	10,000 yr	100,000 yr	1,000,000 yr
Fission Products	0.02	0.02	0.01	—
Actinides	55	14	1.1	0.4
Total	55.02	14.02	1.11	0.4

[a]Fuel burnup of 33,000 MWd/metric ton. d = day; yr = year.
Source: Bell (1973: table A–IV–1).

Reprocessing

The uranium–238 and residual uranium–235 and plutonium in spent fuel can be extracted by what is known as the "Purex" process. This chemical reprocessing technology, developed by the U.S. government in the early 1950s, was initially designed to produce plutonium in a form pure enough for use in nuclear weapons (Metz, 1977: 43). The process was later modified for commercial purposes to allow extraction of uranium from spent reactor fuel.

The spent fuel is chopped into small pieces by an hydraulic shear, exposing the highly radioactive material inside the zirconium alloy jacket ("cladding"). The pieces are dropped into tanks of nitric acid, designed in such a way that a critical mass of plutonium cannot build up at any location. The nitric acid dissolves the fuel, leaving behind the metal cladding. At this point, plutonium, uranium, transuranics, and highly radioactive fission products are all present in solution. The solution is then mixed with an organic solvent. Because of the chemical properties of the solvent, the uranium and plutonium can be extracted, leaving the fission products in solution. Repeated treatment in this manner removes all but about 0.5 percent of the uranium and plutonium in the solution (Lash, Bryson, and Cotton, 1974: 6). The highly radioactive waste solution—the reprocessing "waste stream"—with activities of up to 10,000 curies per gallon, is pumped into stainless steel storage tanks to await disposal.

The fuel cladding becomes very radioactive from irradiation during reactor operation, and enough of the fission products and transuranics adhere to the cladding during reprocessing so that it must be handled as high level solid waste. Significant amounts of radioactive noble gas fission products, such as krypton-85, are normally released into the environment during reprocessing. The Environmental Protection Agency has estimated that the release of all such gases produced in domestic commercial reactors could, by the year 2000, cause a cumulative total of 7,000 adverse health effects around the world, with a 67 percent mortality rate among those affected (Lash, Bryson, and Cotton, 1974: 7).

Other risks are also attendant on the reprocessing of spent fuel. Reprocessing by the international community in order to extract plutonium would result in the accumulation and transportation of very large quantities of this easily fissioned element. Reactor grade plutonium has a critical mass of about twenty pounds—a weight smaller in volume than a grapefruit. Although slightly radioactive, small amounts can be handled with minimal precautions. It is possible that nuclear explosives could be fabricated from reactor grade plutonium by presently nonnuclear nations or terrorist groups, a highly unwelcome possibility. However, a level of protection of plutonium stocks adequate to prevent illicit diversion and use could lead to the abrogation of civil liberties and the establishment of many of the elements of a police state in the United States and abroad (Ayres, 1975). With all of these risks in mind, in April 1977 President Carter announced that spent fuel reprocessing in the United States would be indefinitely deferred with the expectation that other nations might follow the lead of the United States and forego the development of an international plutonium economy with its concomitant transfers of large quantities of plutonium and dangers of nuclear proliferation.

Some other problems, technical and economic, have also contributed to the deferral of commercial reprocessing. The Purex process was designed explicitly to treat "low burnup" fuel whose residence time in a reactor was limited to several months. By contrast, the fuel in light water reactors is being irradiated some five to ten times longer, resulting in higher radioactivity and decay heat levels. These high radiation levels cause technical problems by degrading organic solvents and making maintenance of equipment a difficult proposition.

Attempts to develop alternative reprocessing technologies more suited to commercial reactor fuels have not been successful (see Chapter 4). Moreover, it now appears that commercial reprocessing may not represent an economic benefit. New licensing requirements for existing and planned plants could easily raise the cost of reprocessing to a level that would negate the supposed economic benefits. The arguments on the economics question are, for the most part, too subtle to detail here, but they suggest that, at least for the foreseeable future, reprocessing may not be worthwhile economically.[2]

Since beginning large-scale production of nuclear weapons, the federal government has reprocessed spent fuel from nonpower reactors in order to extract the plutonium needed for building these weapons. In fact, this activity continues even today, despite the deferral of commercial fuel reprocessing. At the present time, three plutonium production reactors are in operation on a twenty-four hour a day, five day a week basis at the Savannah River, South Carolina, plant. The Department of Energy has announced that it intends to reactivate the reprocessing facility at the Hanford, Washington, Reservation in 1980. Throughout the past thirty years, fuel reprocessing for plutonium production has been the source of a vast quantity of high level liquid radioactive waste. In the absence of a permanent disposal method, the liquids have been placed in "temporary" storage tanks (Figure 2–8).

Spent Fuel and Waste Storage

In lieu of reprocessing, spent fuel is now routinely being kept in on-site storage pools because almost no off-site storage capacity is available in the United States. The spent fuel pools of all reactors now in operation were constructed to hold no more than one and one-third to one and two-thirds of a reactor fuel load—that is, sufficient space to store the annual spent fuel discharge plus an entire core load should core removal be required in the event of reactor repair. Thus, after four or five years of reactor operation, the spent fuel pool becomes full. Redesigning the fuel assembly configuration in a pool can increase storage space to a maximum of nine years' annual discharge, but despite the increased storage capacity, studies by industry and government show that as many as twenty-eight nuclear power plants could be forced to shut down by 1986 owing to the absence of storage capacity for additional spent fuel (CERCDC, 1978a: 98).

The quantities of waste presently in temporary storage are quite large (Table 2–4). Liquid wastes from reprocessing for defense purposes total some 75 million gallons (Krugmann and von Hippel, 1977: 884), or roughly 9.5 million cubic feet (IRG, 1978: D–19),

---

2. Despite this, however, the owners of a reprocessing plant at Barnwell, South Carolina, have indicated that the plant will operate if allowed to do so.

Figure 2–8. High Level Radioactive Waste Storage Tanks Under Construction at the Savannah River, South Carolina, Defense Reprocessing Facility.

Source: E.I. DuPont DeNemours and Co.

Table 2-4. High Level Waste and Spent Fuel Storage in the United States.

Existing high level reprocessing waste (as of 10/1/77):

Site DOE-operated		Volume (thousands of cu. ft.[a])	
Savannah River, South Carolina		2900	
Idaho Falls, Idaho (Idaho National Engineering Laboratory)		404	
Hanford, Washington		6102.5	
	Subtotal	9406.5	
West Valley, New York *(Nuclear Fuel Services)*			
Neutralized (Purex waste)		80.2	(600,000 gallons)
Acidic (Thorex waste)		1.6	(12,000 gallons)
	Subtotal	81.8	
	Total	9488.3	

Spent fuel storage (as of 12/31/78):

1. At the end of 1978, there were approximately 4,400 metric tons of commercial spent reactor fuel in storage at reactor sites and the three nonfunctioning reprocessing plants (GE–Morris; West Valley, New York; AGNS–Barnwell).[b]

2. The Tennessee Valley Authority has announced tentative plans to construct a large away from reactor spent fuel storage facility by 1984. The facility may be located at Oak Ridge, Tennessee.

3. The federal government intends to construct an away from reactor spent fuel storage facility by 1984–1985. Interim storage of spent fuel may take place at the three nonfunctioning reprocessing plants.

4. Spent fuel storage experiments are planned for Hanford, Washington, and the Nevada Test Site for 1979–1980. The Waste Isolation Pilot Plant may include "demonstration" storage of up to 1,000 spent fuel assemblies in a retrievable mode, but the project will not commence before 1985.

5. Spent reactor fuel is presently accumulating at the rate of about 1,300 metric tons per year.

[a]1 cubic foot = approximately 7.5 gallons of liquid; much of the defense waste is solidified.

[b]1 metric ton = 2,200 pounds. Volume is about 13.1 cubic feet per metric ton of spent fuel.

Source: IRG (1978).

stored at the government reservations at Hanford, Idaho Falls, and Savannah River. At the end of 1978, about 4,400 metric tons of spent reactor fuel were in storage at reactor sites (Figure 2–9), and the Department of Energy has estimated that, on the basis of a minimum of 148 GWe of nuclear electrical generating capacity at century's end—about three times current nuclear capacity—the cumulative spent fuel discharge by domestic light water reactors will total

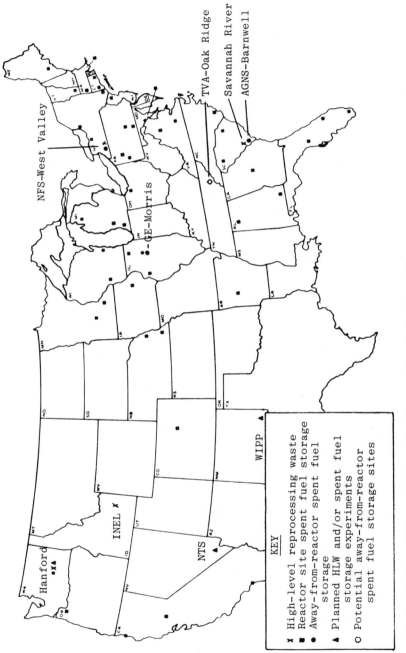

Figure 2–9. High Level Waste and Spent Fuel Storage Sites in the United States.

Source: Prepared by author from DOE (1979c:4) and IRG (1978).

some 70,000 metric tons by the year 2000 (IRG, 1978: D–28). If this spent fuel were to be reprocessed, and one conservatively assumes that the reprocessing of 1 kilogram of fuel produces one-half gallon of liquid, a minimum of 35 million additional gallons of highly radioactive liquid waste would require storage and disposal. (In fact, reprocessing would produce larger volumes of liquid waste.) Assuming an average age of ten years for the reprocessed fuel, the cumulative radioactivity in the liquid waste would be on the order of 21,000 megacuries, including 8,400 megacuries of strontium-90 and 10,200 megacuries of cesium–137 (Hollocher, 1975: 227).

The volume of the unreprocessed spent fuel, in comparison to the defense wastes, is not very great. In terms of radioactivity, however, the two inventories are now of the same order of magnitude. Because the inventory of spent fuel is rapidly increasing at the rate of about 1,300 metric tons annually (IRG, 1978: 227), the radioactivity of the spent fuel already far exceeds that of the defense wastes. As a measurement of comparison of these two inventories, Princeton scientists Krugmann and Von Hippel (1977) calculated the radioactivity of stored strontium-90, which has a half-life of twenty-eight years. Between a few years and a few hundred years after removal from the reactor, the hazard potential of high level waste is dominated by this radionuclide. The number of strontium-90 atoms produced in the fission process is roughly proportional to the total amount of fission energy released in fuel irradiation and therefore provides a convenient benchmark for a comparison of the activities of the liquids and spent fuel. On the other hand, if one uses volume as a measure of comparison, the result is misleading, because the military wastes are relatively "dilute" in comparison to the civilian wastes. Thus, on the basis of strontium-90 inventory, as of 1975, the activity of the defense waste was approximately 270 megacuries, while that of the stored spent fuel was 200 megacuries. By 1980, the activity of the defense waste will have increased only by 15 megacuries, while, assuming a nuclear electric generating capacity of 80 GWe, the civilian inventory will contain some 800 megacuries of strontium-90 activity (Figure 2–10) (Krugmann and von Hippel, 1977: 884). Thus, at the present time, the radioactivity of domestic high level civilian waste far exceeds that of stored high level military waste.

Reactor Shutdown

After an expected operating lifetime of thirty to forty years, a nuclear reactor will be shut down, cleansed of residual radioactivity, and dismantled. This procedure is called "decontamination and de-

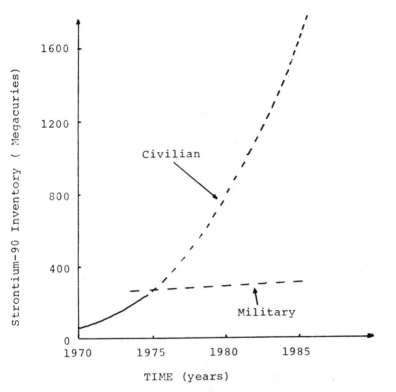

Figure 2—10. Estimated Civilian and Military High Level Waste Inventories Measured in Terms of Their Strontium-90 Content as a Function of Time.

Source: Krugmann and von Hippel (1977:884).

commissioning." Until now, no large light water reactors have been decommissioned. Experience has been limited to the decommissioning of a few small, mildly radioactive experimental reactors. Whether commercial reactors can be decommissioned in the same way is not clear, and cost estimates of the procedure are highly speculative. A Department of Energy report released in February 1978 (hereafter referred to as the "Deutch Report") describes three decommissioning alternatives under consideration—mothballing, in-place entombment, and dismantling-removal (DOE 1978a: 77).

Mothballing consists of removing all fuel and radioactive fluids and wastes and placing the facility in "protective storage"—that is, under lock and key. Entombment includes the above plus shipment of selected reactor components off site, followed by sealing of all remaining highly radioactive or contaminated components within a closed structure. Dismantling-removal requires removal from the site of all materials (including soil) having activities above the guidelines established by the Nuclear Regulatory Commission, leaving the facility owner with unrestricted use of the site.

At shutdown, after forty years of operation, it is estimated that the total inventory of radioactivity in the structure of a typical light water reactor will be 15 megacuries. The reactor vessel, for example, will be intensely radioactive. This could pose difficulties if immediate dismantling is required. The Deutch Report states:

> Because of the high contact dose rate shortly after shutdown, any removal of vessel internals for an LWR would require sophisticated underwater cutting and handling equipment. A delay period of about 100 years would permit sufficient decay of the Co[balt]-60 to allow manual removal techniques with local personnel shielding. (DOE, 1978a: 78)

Therefore, once a reactor is shut down, mothballing for a specified period may well be the preferred, although not necessarily a desirable, method of decommissioning.

Decontamination and decommissioning of a 1,200 MWe light water reactor will produce the following estimated quantities of low level waste (in cubic feet): mothballing—2,000; entombment—70,000; dismantling-removal—500,000 (DOE, 1978a: 84). The Deutch Report estimates the cost of dismantling-removal to be from $27 to $31 million (1975 dollars) (DOE, 1978a: 82); however, a recent inquiry into the costs of nuclear power by the Committee on Government Operations of the House of Representatives found that:

> Dismantling a nuclear plant now may cost anywhere from $31 million to more than $100 million in 1977 dollars—between 3 percent and 10 percent of the $1 billion capital cost. Even the higher figures, however, do not include perpetual care costs for rubble from the plant containing radioactive nickel which may remain hazardous for up to 1.5 million years. After 30 to 40 years, the expected lifetime of a nuclear plant, decommissioning costs would quadruple (assuming 5 percent annual inflation). (House of Representatives, 1978: 22)

There exists some evidence that official cost estimates are much too low. Relatively simple repair and maintenance procedures that have to be carried out in the high radiation areas of today's nuclear plants—for example, replacing sections of piping—already cost millions of dollars due to the difficulties of working in such environments.

The Commonwealth Edison Company of Chicago, for example, is currently in the process of decontaminating its Dresden-1 reactor, a small, 200 megawatt plant brought on line in 1960. The plant is being cleaned in an effort to reduce worker exposure to radiation over the estimated twenty years of useful life left to the plant. Dresden-1 cost $18 million to build and is expected to cost $36 million to clean (*Nucleonics Week*, 1979f: 9). Adjusted for inflation (assuming 5 percent annually over a twenty year period), cleanup costs will amount to some 75 percent of the plant's construction cost. The cost of full-scale plant decontamination and dismantling, using this example as a basis, could easily run into hundreds of millions of dollars.

Cleanup and repair of the Three Mile Island-2 nuclear reactor, which was seriously damaged and contaminated during the course of the accident that began on March 28, 1979, may also provide some indication of the costs associated with decontamination and dismantling of a large power reactor. To be sure, the Three Mile Island plant will not be taken apart, and radiation levels inside the reactor containment structure are considerably higher than those likely to be found in a decommissioned power reactor. Nonetheless, equipment and techniques that will be used to clean and repair Three Mile Island-2 will be similar to those that will be involved in dismantling a shut down plant. For example, the reactor core is apparently a total loss and will have to be removed using remote handling equipment. Monitoring instruments, burnt out by high levels of radiation, and some cooling components, such as liquid seals seriously damaged by radiation, will also have to be replaced. Part of the containment structure and floor contaminated to unsafe levels may have to be removed by drilling or sandblasting. Areas contaminated to lesser levels will have to be washed down to remove residual radioactivity. Solid wastes will have to be packaged for shipment; liquid wastes will have to be solidified. In the absence of a final disposal method, highly radioactive components such as the core may have to be stored in some as yet undeveloped cooling facility. Three Mile Island has been described as a "laboratory" for testing out heretofore untried and untested ideas. Indeed, the cleanup and repair of the damaged reactor may well demonstrate whether or not decontamination and dismantling of a large power reactor is even practical.[3]

3. According to Bechtel, the cost of rehabilitating the Three Mile Island plant will run to about $425 million. This may be considered a conservative estimate (*Nucleonics Week* 1979m: 10).

Considerable quantities of radioactive waste are produced at each step of the nuclear fuel cycle; however, management of the waste generated at certain points of the fuel cycle is more problematic than at other points because of large volume or high radioactivity. For example, the milling of uranium ore produces enormous volumes of sandy tailings with very low levels of radioactivity. These tailings are hazardous because they are easily dispersed and are found, in many instances, in close proximity to human habitations. The day-to-day operation of a nuclear power reactor generates substantial quantities of slightly radioactive trash that cannot simply be dumped or burned but must be packaged and shipped to appropriate disposal sites. Finally, the fissioning of uranium within the reactor creates materials with extremely high levels of radioactivity whose longevity requires that the material be isolated from the environment for extended periods of time. The spent reactor fuel is not particularly voluminous. However, if it is reprocessed to extract residual uranium and plutonium, large volumes of highly radioactive liquids are produced that must be quarantined, at a minimum, for several centuries. The biological hazards of radioactivity require that these wastes be handled with utmost care and in an appropriate manner. In the following chapter we turn to the question of how radioactive wastes can be safely managed, stored, and isolated from the biosphere.

❋   *Chapter 3*

The Management, Storage, and Disposal of Radioactive Wastes

Large quantities of nuclear wastes have already been generated. They continue to be produced, and it is an unpleasant fact that creating the wastes is a good deal easier than disposing of them. The wastes constitute a considerable hazard and cannot be turned loose in the environment—that is, used as landfill or dumped into rivers or oceans. They must somehow be isolated from significant contact with the biosphere.

For the transuranic wastes, the time involved is 10,000 years or more. No human societies have remained in existence for such periods. Because mankind cannot be depended upon to prevent waste under its guardianship from entering the environment, it is necessary to develop a method of isolating the waste so that its safekeeping is independent of human actions and control. Society is not accustomed to thinking about technical challenges involving periods that dwarf the span of recorded history, yet safe isolation of radioactive wastes poses exactly such a challenge.

One solution that has been proposed is the construction of large, intrusion-proof "mausolea," into which the radioactive wastes would be placed and sealed. Experience with such structures, however, indicates that they are unlikely to remain intact for significant periods of time, as with the case of the Egyptian Pyramids. Alvin Weinberg (1972: 34) and others have suggested that surface storage facilities watched over by a nuclear "priesthood" could provide the required millenia of isolation. Once again, however, the continuity of the scheme cannot be guaranteed. Safe disposal of radioactive wastes requires the development of long-lived engineered technologies that will not require human guardianship but will effectively isolate the wastes from human intrusion. These requirements limit the consideration of disposal sites to those parts of the accessible universe that are entirely or almost entirely isolated from the environment and will remain so for a large fraction of the required period. A corollary set of problems encompasses the difficulties of preparing the radioactive materials and the waste disposal site in acceptable ways. Approaches to the packaging of wastes are discussed in this section.

Wastes have been and continue to be put into storage, all of which is temporary. Storage facilities may be located on or near the earth's surface and require constant surveillance. Disposal of wastes implies a terminal solution. Once wastes are disposed of, they are presumably unretrievable. However, there is always the chance that the disposal technology might prove unsafe or inadequate. Thus, disposal with retrievability implies that, for a limited period, the wastes can be removed from their place of interment, substituting either a better disposal method or further storage. We discuss here both storage and disposal technologies, including surface storage, geological disposal, seabed disposal, space disposal, ice disposal, and partitioning and transmutation.

PACKAGING RADIOACTIVE WASTES: MATRIXES AND CANISTERS

The physical form of high level wastes is an important aspect of the waste management program. As will be seen in the discussion of geological disposal, waste form is of critical importance for the emplacement (and retrieval, if necessary) phase of a repository. Over the longer term, waste form may or may not be important, depending upon the degree of sophistication of the waste packaging and chemical conditions within the repository rock matrix.

Solidification of High Level Liquid Waste

Federal regulations require that liquid reprocessing waste be solidified for disposal within five years of production. Several different approaches to solidification have been developed—calcination, vitrification, and incorporation of waste into crystalline ceramics and synthetic minerals.

Calcination is a process in which the liquid waste is sprayed through an atomizer and dried at high temperatures. The resulting granular product, or "calcine" (which is highly radioactive), can then be temporarily stored in bins to await further processing. This practice has been carried out on an extensive basis at the Idaho Falls Reservation (Dickey, Hogg, and Berreth, 1978).

Vitrification involves the mixing of calcined waste with a borosilicate glass frit. The mixture is melted in a special furnace, and the liquid glass may be cast into a mold (Figure 3–1) or poured directly into a metal canister. Vitrification has been under development both in the United States and abroad for nearly twenty-five years. An industrial vitrification plant is in operation on a routine basis at Marcoule, France. In the U.S., the process is still in the experimental stage; the first block of vitrified commercial waste was produced at the Battelle Pacific Northwest Laboratories in early 1979 (Kerr, 1979a: 289).

Borosilicate glass has been considered a suitable matrix for high level waste because the glass, being an amorphous solid with strong interatomic bonding but no strict atomic structure, is able to contain a variety of different elements. At moderate temperatures and pressures the glass has low leachability—that is, under the action of running or standing water, radioactive waste products leak out of the glass at a very slow rate. Furthermore, the glass is resistant to structural damage from radiation (primarily alpha particles) and was, until recently, believed to be fairly resistant to devitrification—that is, cracking and crumbling. Finally, as noted above, the technology to produce large radioactive blocks of glass does exist.

However, some of these supposed advantages have been opened to question. In a series of experiments at Pennsylvania State University in 1978, it was found that under severe temperature and pressure conditions approximating those that might be expected in a granite, basalt, or salt repository, small samples of borosilicate glass containing "synthetic wastes" (nonradioactive isotopes of radioactive waste atoms) placed in distilled water or brine devitrified in a matter of weeks. Large quantities of the synthetic waste escaped into solution; some formed new mineral species not previously present (McCarthy et al., 1978: 216).

These results came as a surprise to many researchers in the waste solidification field, inasmuch as most previous devitrification research was performed at ordinary temperature and pressure. Equally surprising was the fact that some of the new minerals formed in solution actually proved to be more stable and insoluble than the original glass, suggesting that the formation of such mineral species may be one approach to achieving mineralogical equilibrium between radioactive waste and host rock. Indeed, this approach is being studied by groups of scientists in the United States and in Australia as described below.

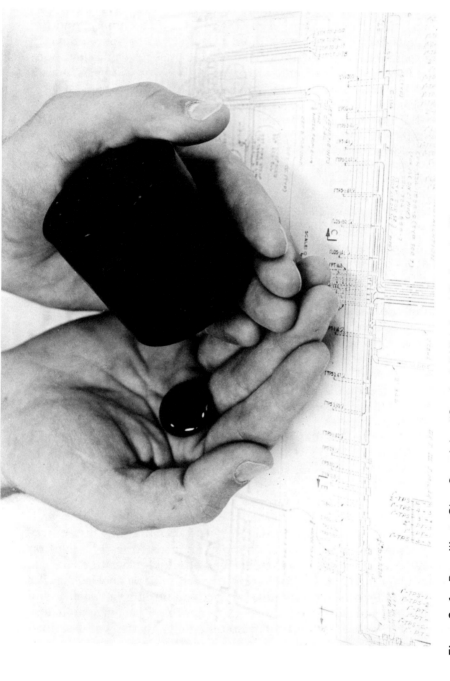

Figure 3-1.   Borosilicate Glass Containing "Synthetic" High Level Radioactive Waste.

Source: DOE.

Crystalline ceramics form another candidate group of possible encapsulants. The ceramic matrix is generally a substance that crystallizes into an ordered atomic structure and that can be tailored to specific waste elements and geochemical conditions. Leachability can be low, and the crystalline structure continues to exist even if the ceramic breaks down. Ceramic containments are currently being studied on a laboratory basis; two such materials are described here.

Supercalcine-ceramic is under development at Pennsylvania State University. Certain elements are added to liquid wastes, which are then calcined, heated, and compressed. The radioactive waste makes up about 80 percent of the ceramic material. During this process, specific minerals with low solubility are formed. Under severe temperature and pressure conditions, supercalcine-ceramic in distilled water shows very low leach rates. However, when placed in brines, the ceramic releases large amounts of waste into solution (Kerr, 1979a: 290; see also McCarthy, 1978b).

Synthetic rock, or synroc, is under development at the Australian National University, where the approach is to find minerals that have proven to remain geochemically stable over geologic time periods and then combine these minerals with specific waste elements. Synroc is only about 10 percent radioactive waste by weight, and in addition, the wastes must be partitioned—that is, chemically separated into groups of elements—in order to attain maximum geochemical stability. Once again, although synroc withstands extreme conditions in distilled water, it does not do so well in brines. The synthetic rock approach appears very promising, but it is likely to be quite expensive as a result of the waste partitioning requirements (Kerr, 1979a: 230; Ringwood et al., 1979).

Ceramics also have several disadvantages. They are subject to damage when the ordered crystalline structure is disturbed by radiation. Another problem is transmutation. When an atom of waste decays, its daughter may have different chemical properties, thus altering the stability of the ceramic. Nonetheless, ceramics appear to be a potentially promising approach to waste solidification.

The time required to develop these new materials is expected to be lengthy. One researcher in the field has expressed the sentiment that for near-term waste disposal purposes, only borosilicate glass technology will be sufficiently developed for large-scale application (McCarthy, 1978a).

Spent Reactor Fuel

Present federal policy calls for the disposal of spent reactor fuel as waste, but extensive research into the disposal of this waste form has not yet been conducted. Exact details are therefore lacking. It is likely that spent fuel will be encapsulated in some type of solid matrix material and immobilized in a metal canister. Some problems could arise with this approach. The degradation of fuel rods in the reactor environment and the continued diffusion of fission product gases out of the fuel could make packaging problematic.

Geochemical reactions are also of concern. The resistance of spent fuel to groundwater leaching is not well known. Researchers at Penn State have found that when reactor fuel is exposed to basalt under extreme temperature and pressure conditions, 100 percent of the cesium in the uranium fuel escapes into solution. However, almost 99 percent of the cesium recombines with minerals in the basalt to form a new, highly insoluble mineral. The basalt also prevents dissolution of uranium (Kerr, 1979a: 291). Nonetheless, a repository will not consist of reactor fuel and rock alone. Therefore, chemical reactions between the spent fuel pellets, cladding, canister, and host rock must be determined. If it is found that the spent fuel breaks down over long periods of time, the probability of accidental chain reactions developing must also be assessed. It has been suggested that gradual geochemical redistribution of fissile nuclides and the decay of nonfissile into fissile nuclides could result in the development of zones of high fissile nuclide concentration. With water inevitably present, criticality might occur (Winchester, 1978a: 2). These questions indicate that spent fuel disposal will not be a simple matter.

Waste Canisters

It has long been assumed that spent fuel rods and solidified high level wastes would be packaged in stainless steel canisters. More recently, it has been recognized that corrosive processes in an underground repository are likely to destroy steel containers within a matter of decades. In response to this, research into the development of corrosion-resistant metals, such as Hastelloy-C and alumina, is currently underway. The coating of canisters with inert metals or glassy materials and ceramics has also been suggested. A recent Swedish study (KBS, 1978) proposed that steel clad vitrified waste packages, after being allowed to cool for forty to fifty years, be encapsulated in layers of lead (in order to reduce emission of radiation that could cause radiolytic decomposition of water and the generation of hydrogen gas) and inert titanium. For spent fuel, this study proposed a container made of copper and ceramic materials. After emplacement in a repository, the canisters would be surrounded with a mixture of quartz sand and bentonite. The sand would have high ionic retention for many radionuclides, and the bentonite, upon contact with water, would expand into an impermeable mass and effectively seal the hole[1] (Barnaby, 1978: 6). None of the canister materials under study are totally corrosion-resistant, but they would be expected to last longer than stainless steel. Even so, a waste canister made of special materials could only be depended upon as a barrier to waste movement during a period of several hundred to perhaps 1,000 years at most (DOE, 1979a: 3.1.59).

Reprocessing

The nuclear industry has long advocated reprocessing of spent reactor fuel as an aid to waste management, claiming that reprocessing reduces the radiological risks associated with radioactive waste disposal by reducing the amount of long-lived radioactivity in and the volume of the waste. However, analyses of the matter have shown these claims to be, for the most part, exaggerated. The American Physical Society's Study Group on the Nuclear Fuel Cycle and Waste Management concluded that:

> Reprocessing is not an essential step in the management of nuclear wastes but rather a means of extending fuel resources. . . . Reprocessed high-level waste has lower actinide content than spent fuel, but miscellaneous transuranic waste is created as a result of reprocessing and [fuel] fabrication; on balance the two waste disposal situations are comparable. (APS, 1977: S8)

This same conclusion was reached by the Ford–MITRE study (1977: 248), the Deutch Report (DOE, 1978a: 2), and the report of the Interagency Review Group on Nuclear Waste Management (IRG 1978: vii).

Reprocessing would not reduce the volume of waste requiring disposal but rather would produce transuranium-contaminated waste with a volume one to three times that of the original spent fuel (Metz, 1977: 45). Reprocessing would not substantially affect the heat output of radioactive waste. Over the short term, reprocessed high level waste would produce as much heat as spent fuel, while over the long term, the lower heat production of high level waste might reduce repository space requirements by only 20 percent (UCS, 1978: 3). In terms of radiological hazard, a significant reduc-

---

1. These proposals have not, however, been experimentally tested. They have been the subject of detailed critiques and criticism. See CERCDC (1978b), Winchester (1978b), and Johansson and Steen (1978).

tion in risk from high level waste as a result of reprocessing would occur only after a period of several thousand years. The near-term risks associated with the handling of extracted plutonium and high level liquid waste would greatly exceed those associated with spent fuel management (Ford–MITRE, 1977: 34). Thus, it does not appear that reprocessing offers any advantage in the management and disposal of radioactive waste.

WASTE STORAGE AND DISPOSAL TECHNOLOGIES

Various solutions have been proposed for waste storage and disposal, limited only by their proponents' imagination. The more serious proposals have been subject to extensive study, and a number have been found to be infeasible, unsafe, or uneconomic. Others are considered very promising. The most prominent proposals are described below.

Surface Storage Technologies

It is expected that by 1983, many reactors will run out of space in their spent fuel storage pools to temporarily store their entire core inventory of fuel in the event of emergency or extensive repairs requiring core removal. Although it would seem logical to expand on-site spent fuel storage capacity to alleviate the space problem, reactor operators are reluctant to make the required investment and also do not want indefinite responsibility for the spent fuel in the absence of reprocessing. It may also be necessary to store spent fuel or packaged high level waste until such time as its heat output declines to a level acceptable for permanent disposal. For these reasons, development and construction of some type of surface waste storage facility seems inevitable. Two such facilities are described briefly here.

Spent Unreprocessed Fuel Facility. The Spent Unreprocessed Fuel Facility (SURFF) was proposed by the Energy Research and Development Administration (ERDA) in 1977. SURFF would consist of a large number of concrete air- or water-cooled vaults into which spent reactor fuel would be placed for periods of up to one hundred years. Several thousand spent fuel vaults would be located at a single SURFF site.

SURFF was not a new idea when it was proposed, however, for it bore more than a superficial resemblance to an earlier proposal called the Retrievable Surface Storage Facility (RSSF). RSSF was a temporary storage facility for high level wastes; the idea was withdrawn shortly after being proposed (see Chapter 4).

Despite the similarities between the two proposals, an interesting semantic difference is perceived by some of SURFF's proponents: while RSSF was clearly only an interim, stopgap facility for the storage of high level waste, SURFF is seen as an antiproliferation project, offering an alternative to reprocessing. It is also obvious that SURFF keeps the spent fuel in a convenient location should reprocessing ever be resumed. In 1974, the Environmental Protection Agency wrote to the AEC:

> A major concern—the employment of the RSSF concept—is the possibility that economic factors could later dictate utilization of the facility as a permanent repository, contrary to the stated intent to make the RSSF interim in nature. (JPL, 1977: 6–44)

The same concerns are applicable to SURFF. In any case, the SURFF proposal is no longer viewed as a serious storage alternative for the near term (see JPL, 1977).

Away From Reactor Storage. The away from reactor (AFR) spent fuel storage pool has been proposed as an alternative to expanded on-site spent fuel storage. The AFR is merely an extremely large spent fuel storage pool—not unlike those found at reactor sites (Figure 3–2)—with a capacity of 5,000 or more metric tons. (By contrast, a typical on-site pool holds no more than 150 to 300 tons of spent fuel.) Construction of an AFR would present no technical obstacles, and the Department of Energy has proposed that one be built by 1984 (DOE, 1978b). Possible sites already in existence for use as an AFR are at the reprocessing plants at Morris, Illinois; Barnwell, South Carolina; and West Valley, New York. Oak Ridge, Tennessee has also been suggested as an AFR site. Use of AFRs for temporary storage would require increased transportation of spent fuel, which has been cited as an objection to such facilities.

Furthermore, as with the RSSF and SURFF, an AFR could become an inexpensive "permanent repository." While there exist no fundamental technical problems that would suggest that an AFR might be unsafe as presently conceived, the AFR concept would relieve reactor operators of all responsibility for spent fuel and would act essentially as a stopgap solution to a long-term problem. Such objections, some of them very important, need to be considered in greater detail before an AFR program is implemented (see DOE, 1978b).

Figure 3–2. Spent Fuel Storage Pool at General Electric's Nonoperational Reprocessing Facility in Morris, Illinois. The facility is licensed to store spent fuel and has been suggested as a possible site for away from reactor (AFR) spent fuel storage.

Source: J.E. Westcott, USAEC.

Geologic Isolation

The disposal of radioactive wastes deep within the earth's crust is considered the most promising of the various proposed disposal technologies. It is also the closest to realization. Geologic disposal is attractive because, in principle at least, it appears that wastes could be safely isolated from the biosphere for thousands of years or longer. Furthermore, disposal in mined vaults is generally believed to require only straightforward applications of existing technology. However, geologic isolation will not be easy to implement, and ensuring the long-term isolation of radioactive wastes will not be possible without an extensive research and development program extending, at a minimum, ten to fifteen years into the future. Because geologic isolation is currently the favored disposal technology, this section discusses concepts, advantages, and disadvantages in more detail than with other disposal concepts.

Advantages of Geologic Isolation. Geologic isolation is considered promising for a number of reasons. First, it is possible to locate mineral, rock, or sediment bodies beneath the earth's surface that have remained intact and not subject either to groundwater intrusion or to seismic and tectonic forces over periods of millions of years. For example, because salt is so soluble in water, the existence of a salt bed implies that it has not been in contact with circulating groundwater for many millions of years. Locations at depths greater than 1,500 feet below the surface are likely to remain unaffected by erosion and isolated for the necessary periods, provided there is no human intrusion. Further, much of the required technology already exists, at least with respect to mined vaults. Finally, the "Oklo phenomenon" gives some measure of confidence that radionuclides can be retained in place in a geologic environment over periods of millions of years.

Oklo, in Gabon, Africa, is the site of extensive uranium ore bodies. Some years ago, it was discovered that because of particular groundwater and chemical conditions at a time several billion years ago, a so-called "natural reactor" developed in several places within one ore body. This natural reactor sustained a low power chain reaction for a period of about 750,000 years. Analysis of the ore revealed that radioactive rare earth and actinide nuclides largely remained in or near the reactor zones. On the other hand, it was found that many fission products, including strontium–90 and cesium–137, were lost completely or in very significant amounts (Cowan, 1976: 46). This discovery has provided valuable evidence that, in theory at least, geologic isolation is a valid concept. The Oklo reactor has demonstrated that, given particular geochemical conditions, retention of long-lived radionuclides is possible.

Some caveats are in order, however. First, a high level waste repository will not closely mimic the chemical conditions present in the Oklo ore. Second, because the precise depth of the Oklo reactor cannot be determined, it is not possible to conclude whether the entire inventory of long-lived radionuclides produced by the reactor remained in place (EPA, 1978a: 26). The complexity of a particular hydrogeological system, erosion, seismicity, tectonic instability, and the possibility of human intrusion into a disposal vault must, therefore, all be considered.

Disposal Concepts. A number of geologic disposal concepts have been proposed, including solution-mined cavities, matrixes of drilled holes, hydrofracture emplacement, deep well injection, disposal on isolated islands, superdeep holes, rock melting, and mined vaults (DOE, 1979a: 1.12–1.25; APS, 1977: S118.) Of these, only the last three have been considered promising enough to warrant extensive investigation.

Superdeep holes would involve emplacement of waste canisters in 4,000 foot columns at the bottom of holes some fifteen inches in diameter and five to eight miles in depth. At such extreme depths, there would exist a minimal possibility of disturbance by climatic or surface changes, groundwater transport of waste to the surface, or human intrusion. To date, oil and gas exploration boreholes have reached depths of 20,000 feet, with bottom diameters of six to eight inches. At greater depths, the local pressure of the overlying geological strata is extremely high, and boreholes do not remain open. Presently, the technology to keep a deep borehole open does not exist; however, there is some reason to believe that holes 40,000 feet deep with bottom diameters of fifteen inches will be technically achievable within the foreseeable future. Although many uncertainties about superdeep disposal do exist regarding retrievability, thermal effects, long-term leaching by groundwater, and borehole sealing, it is considered a potentially suitable technology. It is not, however, presently under active study (see APS, 1977; ERDA, 1976; OSTP, 1978b; DOE, 1979a).

Rock melting would take advantage of the high heat production of some forms of high level liquid and solid waste. Certain types of high level waste generate so much heat that, if placed in vaults in or pumped into crystalline or igneous rock, they would cause melting of the surrounding material. After fifty years or so, the temperature of the waste would begin to decline and the rock to solidify (Figure 3–3). In theory, this would permanently immobilize the waste. If

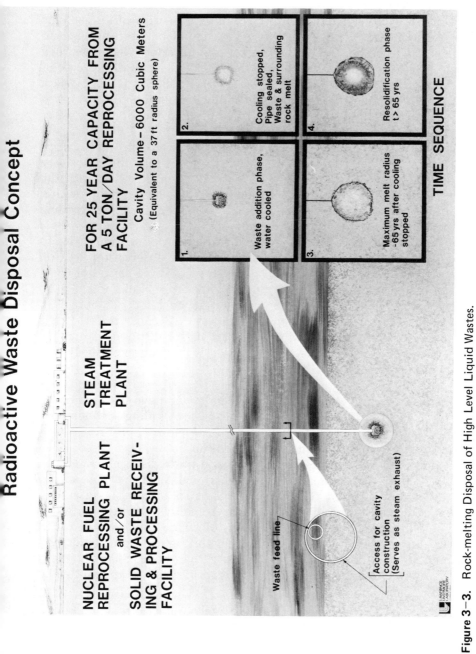

Figure 3–3. Rock-melting Disposal of High Level Liquid Wastes.

Source: DOE (1979a:3.4.2).

located near a reprocessing plant, rock melting could be particularly advantageous, because the liquid waste stream could be pumped directly into the rock stratum. Transportation would not be required, and disposal costs could be low by comparison with other geologic disposal concepts. However, the waste would be irretrievable and subject to leaching and transport by groundwater. Although rock melting is under study, it is viewed only as a long-term disposal alternative (see ERDA, 1976; APS, 1977; OSTP, 1978b; DOE, 1979a).

Disposal in mined vaults, the only geologic isolation concept based on available technology, would involve the construction of a waste repository in some geologic medium at a depth of 1,500 to 2,000 feet (Figure 3—4). Waste canisters would be lowered through a vertical access shaft and emplaced in cavities bored into the repository floor (Figure 3—5). These boreholes would then be filled with the original material or possibly with materials with high ionic retention and impermeability (to water), such as clay or zeolite. The individual rooms within the facility would be backfilled, but the access shaft and tunnels would be left open for a certain period of time—five to twenty-five years has been suggested—to allow observation, and removal if necessary, of waste canisters. Over this short period of time, however, the long-term integrity of the repository could not be decisively determined. Finally, the vault would be filled, all boreholes and shafts sealed, and the facility abandoned. Because mined vaults are considered the most promising locations for disposal of radioactive wastes, the remainder of this section is devoted exclusively to a discussion of this disposal technology.

Disposal Media.   A variety of geologic media have been proposed for the location of geologic repositories, including bedded and domed salt, granite, basalt, shale, and tuff. Figures 3—6 through 3—9 show areas in which these media can be found.

Bedded salt was first recommended as a waste disposal medium in a 1957 report by a committee of the National Academy of Sciences, in which it was noted that because salt is so soluble in water, the presence of salt beds indicates isolation from circulating groundwater and stability over hundreds of millions of years. Salt beds were laid down during the evaporation of prehistoric seas. Domed salt was formed when salt beds of lesser density than adjacent rock "bubbled" upward through the overlying strata (a process known as diapirism).

Salt is plastic and flows under pressure, a property that assures that openings and discontinuities are sealed; it dissipates heat well

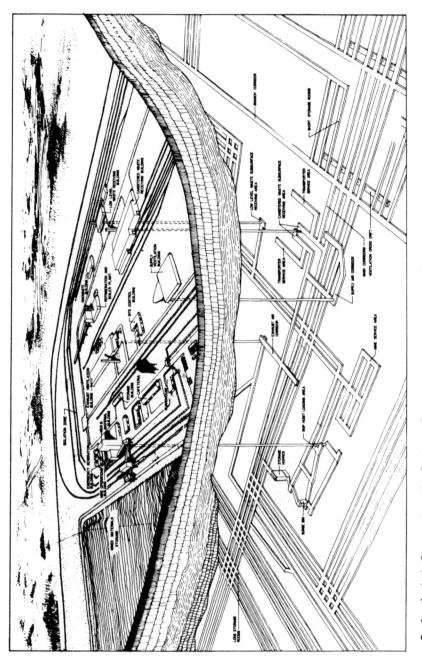

Figure 3–4. Artist's Conception of a Geologic Repository and its Support Facilities.

Source: DOE (1979a:3.1.103).

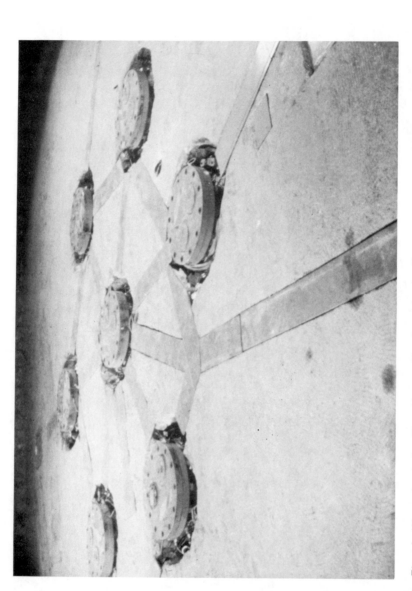

Figure 3–5. Electric Waste-simulating Test Heaters Emplaced in the Floor of the Lyons, Kansas, Salt Mine as Part of Project Salt Vault.

Source: DOE.

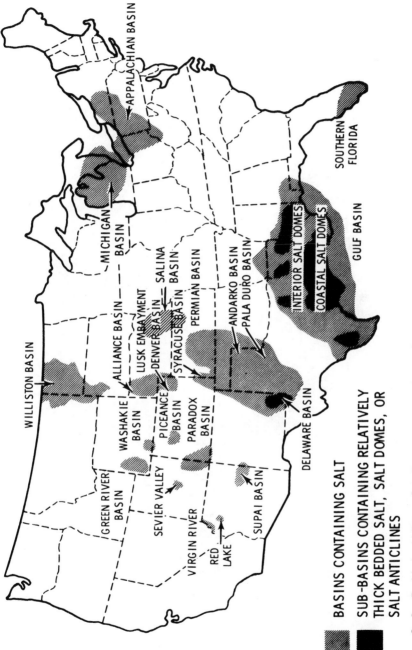

Figure 3–6. Bedded and Dome Salt Deposits in the United States.

Source: DOE (1979a:3.1.10).

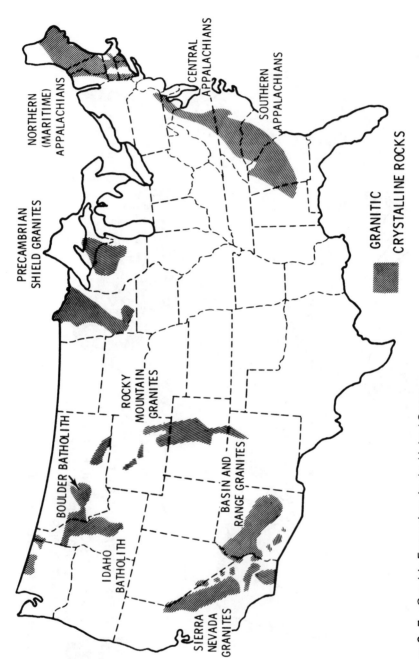

Figure 3–7.   Granitic Formations in the United States.

Source: DOE (1979a:3.1.11).

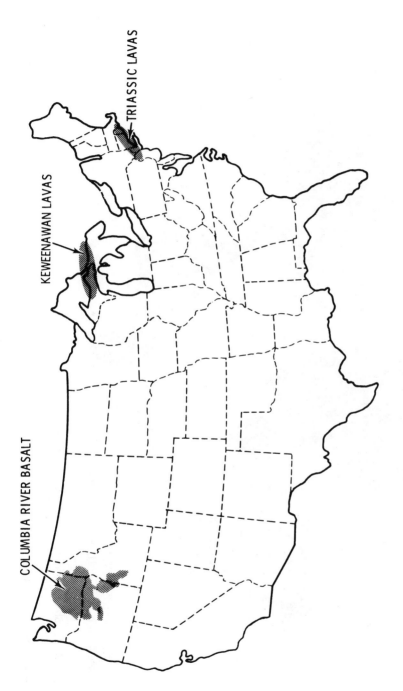

Figure 3–8.   Basalt Formations in the United States.

Source: DOE (1979a: 3.1.14).

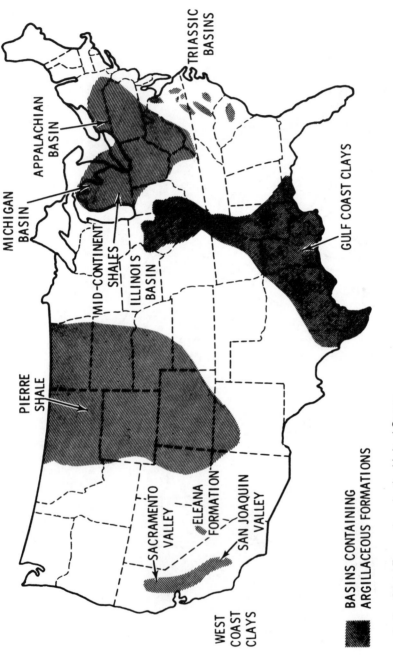

Figure 3—9. Shale Formations in the United States.

Source: DOE (1979a: 3.1.13).

and is highly impermeable to noncirculating water. However, in addition to its high solubility in fresh water, salt contains small quantities of corrosive bittern brine, is weakened by heat, and expands when heated. Salt's plasticity could make waste difficult to retrieve, if this became necessary. Furthermore, salt is frequently associated with oil and gas and is itself a natural resource. Some of the technical uncertainties regarding salt behavior and their relation to radioactive waste disposal are discussed below. (For a discussion of past and current research into salt disposal, see Chapters 4 and 5).

Granite is a strong, coarse, crystalline igneous rock with chemical and structural stability and good heat tolerance. The rock itself has a low water content, but shallow granitic formations are generally riddled with water-filled fractures. However, the high purity and low ionic content of the water suggest that corrosion of waste canisters might not be a problem in granite. Also, at great depths, the extreme local pressures may make granite nearly impermeable to water flow. Granite is presently under investigation in a joint Swedish–U.S. program in an abandoned iron mine in Sweden and in a program run by the Department of Energy at the Nevada Test Site (see Chapter 5).

Basalt, a dense, strong, volcanic rock, exists in relatively thin layers interspersed with clay minerals. It has low permeability and moisture content and remains strong even at elevated temperatures. However, basalt is usually full of joints that are potential channels for water flow. Basalt is currently being studied at the Hanford Reservation (see Chapter 5).

Shale, produced by the compaction and cementation of mud, is found in layers. Shale has low permeability and solubility, high ionic retention, and flows under pressure. It also contains water, which in some instances may be saline, and this could cause canister corrosion. Shale layers generally contain fractures and joints that, below the water table, are full of water. High temperatures may cause weakening, dehydration, and crumbling of shale, although the extent of these effects is not yet clear. Shale is being studied in various federal and university laboratories.

Tuff, an extrusive volcanic rock, exists in two forms. The first, welded tuff, is a high density rock with low porosity and water content. Physically similar to basalt, it generally contains many water-filled fractures. The second form, zeolitic tuff, has low density, high water content, and good ionic retention for important radionuclides. However, as with shale, high temperatures cause dehydration and fracturing of zeolitic tuff and may also cause some decomposition to new minerals. Tuffs are young, geologically active rocks that are subject to faulting. Both types of tuff are being investigated at various government and university laboratories. Researchers from Sandia Laboratories and Los Alamos Scientific Laboratory are studying the feasibility of waste disposal in tuff in southern Nevada (Kerr, 1979b: 605).

Technical Uncertainties about Geologic Isolation.   In this and the following section, some of the technical uncertainties associated with geologic isolation in general and salt disposal in particular are discussed in detail (see also Chapter 5). The early optimism over geologic isolation of radioactive wastes has faded somewhat as more extensive research into problems associated with the technology has taken place. A draft report on gaps in the knowledge base, released by the President's Executive Office of Science and Technology Policy in July 1978, put the problem of ensuring long-term isolation of wastes into perspective:

> The task of accurately predicting the fate of radionuclides emplaced in a repository over time frames of even several hundreds to many thousands of years is unprecedented. The principal earth science disciplines involved in such a task—namely hydrogeology, geochemistry, and rock mechanics— are relatively young sciences with little experience in making or evaluating predictions that cover time frames even as short as a few decades. (OSTP, 1978a: 11)

That paper, and a subsequent draft issued the following October (OSTP, 1978b), identified numerous limitations in the data available on geologic isolation. Several of these limitations are discussed here.

1. *The effects upon the host rock caused by repository construction and burial of thermally hot wastes are poorly understood.*   The establishment of a waste repository in a geologic medium can cause long-term perturbations in the rock and in any groundwater present in the medium. Mechanical disturbances to the host rock are caused by mining. Stresses caused by this operation can induce fracturing. In a previously unfractured medium, such as salt, this could open a pathway for intrusion of groundwater into the repository.

The types of chemical reactions likely to take place between radioactive wastes and the host rock are poorly known, and in particular, the chemistry of many transuranic elements in geologic media is not well understood. These reactions could lead to weakening of the medium in the vicinity of the wastes.

The chief radiohydrologist of the U.S. Geological Survey, George DeBuchannane, has stated:

> The release of thermal energy in the geologic media will result in a variety of mechanical, mineralogical, and hydrologic effects that may strongly influence transport of wastes away from a repository. These effects, some of which are imperfectly understood, are extremely difficult to predict for the intervals of geologic time required for the isolation of the wastes. (1978: 8−9)

Amplifying on this, a Geological Survey Circular about technical uncertainties associated with geologic isolation admitted: "[G]iven the current state of our knowledge, the uncertainties associated with hot wastes that interact chemically and mechanically with the rock and fluid system appear very high" (Bredehoeft et al., 1978: 6).

2. *Mechanisms of groundwater flow into and radionuclide transport out of the repository must be identified.* A suitable geologic site should be essentially free of the circulating groundwater that would constitute the primary transport mechanism for radionuclide migration into the biosphere. A site that is apparently free of groundwater may not always be so, according to the American Physical Society:

> Although a proposed site may at the present time be "dry" and seem free of the effects of groundwater, it undoubtedly is, or at some time during the period of concern (up to one million years) will be, in fact, located within an active groundwater flow system. (APS, 1977: S123)

Furthermore, the science of groundwater flow is still very much in its infancy. Determination of the pattern of water flow through fractured rock is dependent upon precise knowledge of fracture geometry, but the complexity and variability of this geometry is so great as to make impossible—at least for the present—construction of an adequately detailed model of groundwater flow. A recent review of groundwater flow models, quoted in a Geological Survey Circular, noted that "[T]he complete description of the quality and quantity aspects of groundwater flow under complex . . . conditions . . . is still in the future" (Bugliarello and Gunther, 1974, quoted in Bredehoeft et al., 1978: 9).

Groundwater flow may also be affected by climatic changes in the area of the repository. DeBuchannane has pointed out that:

Major climatic oscillations, with periods on the order of tens of thousands of years, have been a feature of global climate for at least the past million years and may be expected to continue. Therefore, existing paleo-climatological data need to be reviewed to judge the likelihood of the wastes being exposed during a future erosion cycle and/or transported as a result of change in the hydrologic regime. (1978: 9)

Predictions about climatic oscillations of this type can be made only with great uncertainty.

The repository must be sealed at the end of the operational period in order to prevent the intrusion of surface water and the formation of short circuits between overlying aquifers and the facility. Once the geologic formation in which the repository is to be located has been breached, the resistance of the medium to intrusion by water is seriously reduced, particularly in the case of salt, which is highly soluble in water. Drill holes and mine shafts both represent potential pathways for water intrusion and radionuclide escape. According to DeBuchannane, "All repository openings, regardless of depth, will have to be sealed by some, as yet unknown, technology which will in effect return the site, as nearly as possible, to its original condition" (1978: 10).

Intensive research is now underway to develop cements and grouts suitable for sealing such a repository. Even materials developed expressly for this purpose may have limited effective service lives. According to the Office of Science and Technology Policy, "Consensus does not now exist that borehole sealing materials and technology . . . will provide an adequate seal *even over time frames of decades*" (emphasis added; OSTP, 1978b: Appendix A, 42).

3. *Geologic behavior of the repository area must be predicted over time spans of hundreds of thousands of years.*   Confident prediction of geologic stability is highly critical to repository siting. It is a difficult task because geology is essentially a retrospective rather than a predictive science. Repositories will, it is hoped, be sited only in areas with no history of geologic instability. Unfortunately, there is no way to be sure that a particular area will remain seismically or tectonically stable in the future. According to a group of French geologists, "The past stability of an area is not sufficient to assess a probability coefficient for the future stability of the same area" (DeMarsily et al., 1977: 521). Furthermore, historical records are

somewhat limited in this regard. As an example, had the very damaging Charleston, South Carolina, earthquake occurred in 1500 rather than in 1886, no record of seismic activity in that area would exist, and the area would be assumed stable. Those areas of North America historically free of earthquakes have not always been thoroughly investigated for past seismicity. Inactive faults in the basement rock are essentially undetectable, but within a million year period may, at some point, become active.

Would an earthquake seriously affect a repository? It is believed that the effects of an earthquake in the general area of an existing mined repository would not be great, for numerous examples can be found of mines that have not been damaged by earthquakes. Yet a fault that developed directly through a repository after closure could provide a short circuit for groundwater intrusion into the facility. A nearby earthquake could have potentially serious effects upon the local groundwater flow patterns. Either event could reduce the long-term effectiveness of the facility by allowing the escape of portions of its radioactive inventory.

4. *The response of waste canisters to emplacement in particular geologic environments and the effects upon canister retrievability must be evaluated.* The longevity of waste canisters, once emplaced in a repository, is unknown. If retrievability is desired, the canisters must remain intact for some minimum period of time. However, metallic containers are subject to chemical attack in most, if not all, geologic environments. A recent report written by an Ad Hoc Panel of Earth Scientists for the Environmental Protection Agency expresses the opinion that "It is unlikely . . . that the integrity of the canister, its contents, and its immediate surroundings will last very long. . . . We have seen no evidence of survivals longer than a decade" (EPA, 1978a: 44).

The report concludes:

> Retrievability of HLW [high level waste] . . . is not so much a question of locating the canisters because they have bodily moved elsewhere, but being able to collect all of the waste because corrosion and leaching might so disintegrate the canisters that much of it is dispersed. (EPA, 1978a: 43)

It has been suggested that use of certain highly inert metal alloys, such as titanium, could eliminate the problem of early canister disintegration, but little research has been performed on this approach.

Recognition of these, and other, uncertainties has prompted a reassessment of basic geologic disposal concepts. It is increasingly believed that, although the waste matrix, canister, and overpack material may provide a significant barrier to water intrusion and radionuclide movement, engineered barriers cannot be solely depended upon for long-term containment of radioactive wastes. Furthermore, it is unlikely that any repository site will remain untouched by groundwater over geologic time periods. As a result, many in the waste management field are suggesting that the efficacy of a particular geologic disposal system must be examined in the context of the entire hydrogeologic system in which the repository is to be located, rather than as a function of specific artificial barriers to radionuclide transport. According to the Office of Science and Technology Policy:

> [E]mphasis is on the *system* comprising the entire hydrogeologic and geochemical environment and on the multiple barriers that it provides. This . . . emphasis leads to a site selection and design philosophy that relies upon a series of backup systems in the event of reduced performance of a key component. . . . A geologic environment combining long groundwater flow paths, slow groundwater velocity, and rocks with high sorptive capacity along the flow path would provide multiple natural barriers. . . . Multiple barriers can also be "engineered" . . . [and] . . . [a]lthough the efficacy of engineered multiple barriers might be shorter-lived than those provided by nature, they could be effective in some geologic environments. . . . (OSTP, 1978b: Appendix A, 22)

Thus, although many uncertainties concerning the behavior of geologic repositories do exist, some of which could pose formidable obstacles to proper repository operation, their existence does not necessarily preclude development of geologic isolation as a practical waste disposal technology.

Technical Uncertainties About Salt Disposal. Although salt has long been considered the prime candidate as a radioactive waste disposal medium, many gaps in technical knowledge remain to be filled before a salt repository can be built and used with confidence. Some of these gaps are discussed below.

1. *The behavior of brine, particularly that found in small inclusions or cavities, must be evaluated to determine its effect on salt integrity and canister longevity.* Bedded salt contains tiny inclusions of salt-laden water called "brine," which may comprise 1 to 2 percent of the salt by weight. Salt becomes more soluble in water if heated, and this property allows brine inclusions to move through salt in the direction of a heat source. Thus, brine pockets located at a distance from a hot waste canister would migrate toward the can-

ister. If the temperature in the salt were sufficiently high, the inclusions might decrepitate, or burst, causing local fracturing of the medium and possibly impairing its integrity. It is believed that brine would collect around each waste canister at the rate of about one to two quarts per year for the first twenty years after emplacement, tapering off to negligible quantities after fifty years (OWI, 1976: 103). In the event that spent fuel was buried, brine inflow might continue for longer periods because the thermal output of the fuel drops off more slowly than that of high level waste.

For a long period of time, the presence of brine in salt was not considered a significant problem, and it was assumed that, in any case, liquid present in the repository would be vaporized and carried away by the ventilation system. However, external canister temperatures may not be sufficiently high to vaporize the brine, and concern has emerged over the possibility of canister dissolution. There are, in effect, two extreme views of the problem. The optimistic view asserts that brine inflow is minimal and that encapsulation of the waste in inert glasses, ceramics, and metal alloys will prevent or minimize the possibility of canister dissolution. Those less sanguine about the brine problem note that in the presence of air, brine is highly corrosive and will attack virtually all metals and metal alloys, except noble metals such as gold. Furthermore, even the steam produced by vaporization will attack metals.

The problem of brine-induced corrosion has been known for some time. It was observed during Project Salt Vault at Lyons, Kansas, an experimental project in salt described in Chapter 5. According to the American Physical Society, after retrieval, canisters and heaters emplaced in salt showed:

> [A]ttack in the salt environment of heater shrouds and simulated waste canisters . . . presumably resulting from the release of included brine. The surface of the . . . canisters and heaters . . . maintained at temperatures well above the boiling point of the brine, though superficially in good condition, showed extensive stress corrosion cracking. . . . Much more severe was the attack of the stainless steel conduit feeding . . . [a] heater at a position surmised to be where condensation of the brine-generated steam was occurring. (APS, 1977: S113)

Carbon steel was less severely corroded than stainless steel. Project Salt Vault lasted only three years and produced limited amounts of data on the brine-induced corrosion problem. Under the conditions expected to exist in a salt repository, a metal waste canister could be expected to dissolve completely within one to two decades, at which time the waste matrix would come into direct contact with the brine and salt. Depending upon the insolubility of the glass or ceramic matrix, further dissolution might or might not occur.

The presence of brine could cause other problems:

a.   According to the report of the Ad Hoc Panel of Earth Scientists, "Because of the high density of the [waste] canister, it might sink in any salt bed that contained . . . saline water solution, being corroded and leached as it moved downward. Many beds of salt overlie permeable limestone formations and the canisters might end up there, in the path of groundwater movement" (EPA, 1978a: 18).

b.   As the metallic waste canister is corroded by the brine, it will form oxide, chloride, and hydroxide corrosion products, and the water in the brine will react to form hydrogen, which is highly explosive if exposed to heat. In the absence of ventilation, ignition of the gas could lead to fracturing of the salt.

c.   Rearrangement of spent fuel canisters and transport of actinides as a result of brine-induced corrosion might lead to the development of zones of fissile elements in concentrations sufficient to sustain low level chain reactions, thereby causing further disruption of the repository.

d.   Vaporization of the brine into steam could lead to salt fracturing if pressure were to build up to sufficiently high levels. Presently, quantitative evaluation of the behavior of gas in rock is impossible, the behavior of a water-steam presence in salt is poorly understood, and the effect of steam upon repository integrity is difficult to evaluate. However, adequate ventilation of the facility over an extended period of time could do much to minimize this problem.

2. *The effects of thermal loading on salt must be determined.* The magnitude of the effect of thermal energy on a salt bed is unclear. When a heat source is placed in a geologic medium, it generates a thermal pulse that propagates through and causes a general elevation of temperature in the medium. In bedded salt, this could be a problem, according to the Office of Science and Technology Policy, because

The thermal expansion of salt is almost three times that of other [rock types] . . . [and] is the major design parameter controlling the vertical uplift and induced stresses resulting from the increase in temperature in the repository and surrounding geologic media. A major effect of the thermally induced uplift is the ... deformation of the surrounding media which have protected the salt formation from ground water. (OSTP, 1978b: Appendix A, 64)

Such deformation could cause fracturing and allow groundwater to enter the repository. The high solubility of salt in water requires that groundwater intrusion be minimized.

The interaction of the hot waste and salt could also cause the canister to rise or sink through the salt bed. This problem is considered serious enough to have generated two separate studies (Hyder, 1977; Dawson and Tillerson, 1978), but there is general disagreement over the degree of movement. Significant canister movement would effectively rule out any possibility of retrieving the waste at a later date.

3. *Mechanisms that might cause subsidence of overlying rock layers must be studied, and problems arising from the plasticity of salt must be determined.* Salt is a plastic mineral—that is, it tends to flow slowly like a very thick liquid. This property is helpful in sealing holes, fractures, and openings that may be present or might develop in the material. After completion of the operational period, the repository will be backfilled with loose salt, which will eventually create an impermeable mass. Although this is an advantage in preventing easy access to the wastes, it could present a problem should rapid access be required after the repository is sealed. The technology to assure such access does not now exist. Further, if salt is used as backfill, it will contract to about 80 percent of its original volume. As the overlying rock layers settle and fill in the openings left by the contracting salt, fracturing of these layers may occur. Once again, these fractures could provide a path for groundwater access to the salt bed (Bredehoeft et al., 1978: 5). The contraction problem might be avoided through the use of materials such as concrete, but these materials could be subject to chemical attack by brine.

4. *The optimal waste form must be determined, and the effect of waste form upon repository integrity must be assessed.* Appropriate chemical and mechanical form of the waste is critical to confidence in the disposal technology. The behavior of some waste forms under conditions of elevated temperature and pressure has been described previously. Also critical is adequately low thermal (heat) output of the waste form. Elevated waste temperatures complicate the quantitative evaluation of waste-mineral interactions and may seriously compromise the integrity of a salt repository (Bredehoeft et al., 1978: 6). Canisters of high level waste could be engineered to pro-

duce a lower thermal output; however, the physical form of spent fuel could present problems in this regard. Wide spacing of spent fuel canisters could reduce the effects of high temperature on the entire repository, but would not eliminate effects in salt adjacent to the canister. On the other hand, if spent fuel assemblies could be stored temporarily for fifty to one hundred years, the heat problem would diminish significantly. According to Geological Survey radiohydrologists, bedded salt is probably an excellent disposal medium for low level and transuranium-contaminated wastes, both of which produce little or no heat, but less suitable for spent reactor fuel that has been cooled for only ten or twenty years (DeBuchannane and Wood, 1978).

Seabed Disposal

Disposal of radioactive waste in the deep seabed is an attractive concept for several reasons: the seabed is remote from human activities, it seems possible that the seabed could provide the required long-term isolation, large areas are available, and, not the least important, seabed disposal could avoid some of the political difficulties of waste disposal on land.

Although the concept has been under active investigation since the mid-1970s, numerous technical uncertainties about seabed disposal remain to be resolved. For example, seabed response to the emplacement of hot wastes and the means whereby radioactive materials might enter into aquatic food chains are unknown. Furthermore, because retrievability of the emplaced waste would be difficult, if not impossible, the potential that exists for uncontrolled diffusion of radioactive waste into the ocean environment is not an attractive one. Nonetheless, if these and other problems can be dealt with, seabed disposal could be a suitable alternative to land-based geologic waste disposal.

Rather than the simple dumping of waste on the ocean floor, as has been widely practiced in the past (and which continues, to some extent, today), seabed disposal consists of emplacing wastes in sub-ocean geologic formations. Three such formations have been proposed as suitable for emplacement—deep ocean trenches, deep ocean sediments, and subsediment bedrock.

Deep ocean trenches, such as are found off the coasts of Alaska and South America, exist as a result of contact between crustal masses or "tectonic plates." At these locations, the oceanic crust is being overriden by the lighter land crustal rock at the rate of one to three inches per year. The ocean crust plunges deep into the earth.

This process, called "subduction," is accompanied by vulcanism and seismic instability. It has been suggested that waste canisters dropped into these trenches would be drawn into the earth's crust and safely isolated. A simple calculation shows, however, that over the period of greatest concern (250,000 years), the crust would only move about ten miles, and during the period of greatest hazard (1,000 years), only some 250 feet. In effect, the waste would be left on the ocean floor, which, for obvious reasons, is undesirable. As a result, this approach is not under serious consideration (ERDA, 1976: 25–33; DOE, 1979a: 3.6.4).

The subsediment bedrock has also been suggested as a potential disposal location, particularly if disposal in sediment proves unsatisfactory. At the present time, however, bedrock disposal is not being investigated (ERDA, 1976: 25–45; DOE, 1979a: 3.6.7).

The deep ocean sediments are found primarily at the "midgyre, midplate" points in the ocean at depths of two to three miles, in the center of the suboceanic tectonic plates. The terrain in these areas is gently rolling "abyssal" hills. The hills are generally covered by 150 to 300 feet of red clay sediment, composed of particles that have been settling to the ocean bottom evenly and without interruption for millions of years. Sediment core samples show that, in some areas, this process has been going on for seventy million years (Kerr, 1979b: 606), and cores taken hundreds of miles apart often show the same sedimentary characteristics. The abyssal hills are seismically and tectonically stable and appear to be suitable locations for emplacement of canisters of radioactive waste. Sediment disposal is presently under active study by scientists at the Woods Hole Oceanographic Institution in Massachusetts. Emplacement of the waste canisters could take one of several forms (Figure 3–10). A canister could be buried about one hundred feet deep in the sediment by means of a free-fall ballistic "penetrometer" (a hydrodynamically stable body with a pointed nose and stabilizing fins). Alternatively, drilling rigs could bore holes in the sediment, into which a canister would be lowered by cable. The latter approach could offer greater potential for canister retrievability.

The red clay sediments under consideration have a number of physical and chemical properties favorable to the isolation of radioactive waste. Like salt, the clay flows under pressure. This plasticity ensures that entry holes would seal up soon after canister emplacement. Although the clay is not impermeable to water, active water flow does not occur, because of the absence of forces that could drive such a flow. Under normal conditions, movement of waste would be by sim-

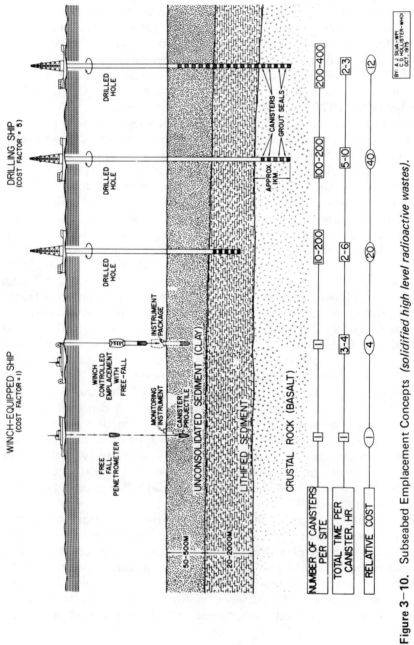

Figure 3-10. Subseabed Emplacement Concepts *(solidified high level radioactive wastes).*

Source: C.D. Hollister, Woods Hole Oceanographic Institution.

ple diffusion alone. Calculations suggest that it would take the waste many tens of thousands of years to diffuse to the sediment surface (Kerr, 1979b: 606). The clay also exhibits strong ionic retention for certain radioisotopes. As a result, in the absence of active transport processes, wastes could theoretically be retained in the sediment for millions of years.

However, a number of questions remain to be answered regarding the ways in which active transport mechanisms might develop within the sediment. For example, the high heat output of the waste during the first one hundred years after emplacement could induce higher water diffusion velocities through the clay. The thermal potential produced by the waste in the porous sediment could cause movement of radionuclides through the clay (APS, 1977: S117). Heat from the waste could also cause a change in the volume and viscosity of the sediment, causing the canister to sink or, more significantly, the sediment to rise toward the water column (DOE, 1979a: 3.6.7). The total effects of heat and radiation upon the sediments are not yet known.

Although it has been suggested that should the waste actually reach the sediment surface, the several miles of ocean water above that surface would provide a barrier to further waste movement, such is not the case. Even at supposedly "calm" seabed sites, small currents have been measured that could carry radioactive atoms across the oceans within periods of several hundred to 1,000 years (Kerr, 1979b: 606). Once on the sediment surface, ion retention for elements such as plutonium is no longer significant (ERDA, 1976: 25.43). The role of aquatic biota in waste transport could also be significant. Recent research has shown that animals on or near the deep sea bottom may actually be part of overlapping food chains that reach all the way to the surface (Kerr, 1979b: 606).

Problems are also expected to arise with regard to waste form. Both spent reactor fuel and vitrified high level waste could be buried in the sediment. However, hot, moist, saline sediments are known to be very hostile environments for both glass and metals. (DOE, however, believes that canisters with lifetimes of 600 years can be developed.) In any case, waste form and canister cannot be depended upon to provide long-term containment. The primary barrier will have to be the sediment.

A major obstacle to seabed disposal is international law. Under the London Convention of 1972, the dumping of high level waste in international waters is expressly forbidden. (The agreement does not, however, ban dumping of low level wastes in territorial waters, practiced extensively by Great Britain and Japan.) The treaty does not address the controlled disposal of high level waste or spent fuel in the seabed, and this point will have to be clarified.

In spite of the many uncertainties, seabed disposal is viewed as a promising alternative to geologic disposal on land. In terms of development time, seabed disposal is some years behind geologic isolation; researchers in the field estimate that the technology cannot be implemented before 1995 at the earliest, even with greater funding and research (OSTP 1978b: Appendix B,7). At the present time, estimated costs are very uncertain, although they are likely to be comparable to or somewhat greater than costs for a geologic disposal system on land.

Finally, should intractable technical and political problems develop with regard to land disposal, seabed disposal could come to be regarded domestically and internationally as an ideal solution. Still, the idea must now be regarded as attractive only in principle, for the potential for ecological disruption of the oceans, however small, does exist and has not yet been demonstrated to be acceptably low (see ERDA, 1976; Anderson, Hollister, and Talbert, 1976; Grimwood and Webb, 1977; APS, 1977; Deese, 1978; OSTP, 1978b; Kerr, 1979b; DOE, 1979a).

Space Disposal

Outer space has long been viewed by many as the ultimate garbage dump. Why not, ask some, rocket radioactive wastes into the sun or even out of the solar system? In fact, although the technical uncertainties of space disposal are great and the costs promise to be high, the notion of permanently eliminating radioactive wastes from the earth is attractive enough to have merited serious and extensive study. However, for reasons discussed below, an environmentally acceptable safe program would only be able to eliminate a fraction of the radioactive wastes requiring disposal. Reprocessing would be required to hold the volume of waste to manageable proportions. A terrestrial disposal system would be required for leftover radioactive materials. A failsafe ejection system or a waste canister able to survive reentry and impact would be required. Despite its attractiveness, space disposal appears to be a dubious proposition at this time.

In the concept currently under study by NASA and the Department of Energy (Figure 3–11), a mass of solidified waste would be placed in a specially designed nuclear waste payload container, loaded into a space shuttle, and lifted into earth orbit. A second shuttle would lift an orbital transfer vehicle (OTV) and a propulsion "kickstage" into orbit. The waste container would be loaded into the front end of the kickstage. The OTV would propel the waste vehicle

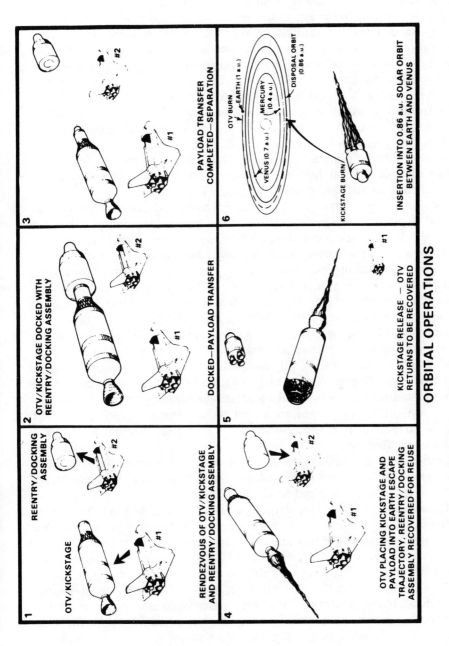

Figure 3–11. Space Disposal Solar Orbit Ejection Sequence.

Source: DOE (1979a:3.10.6).

into an earth escape orbit. The kickstage would then maneuver the waste vehicle into a stable orbit around the sun, about halfway between the orbits of earth and Venus, where the package would remain for at least ten million years. Alternative destinations such as the moon, the sun, or interstellar space have also been suggested. However, location of waste disposal sites on the moon could hinder future lunar exploration. Ejection into the sun or interstellar space requires a very high orbital escape velocity, beyond the capability of currently planned propulsion systems. Waste could also be lifted into a high earth orbit, but such an orbit is not stable over the required time period. Furthermore, a large number of waste containers would have to be placed in orbit. As DOE has somewhat dryly noted, "The idea of several hundred nuclear waste containers orbiting the Earth is not an appealing one" (DOE, 1979a: R.19).

The number of space shuttle flights required for waste disposal would depend on the composition of the waste. Disposal of dissolved spent fuel—excluding cladding and gases—for a nuclear commitment of fifty large light water reactors (about one-quarter of the number in operation, under construction, or planned) would require 600 shuttle flights each year (DOE, 1979a: 3.10.19). These flights would consume the equivalent of almost 23 percent of the electrical energy generated by the reactor fuel (DOE, 1979a: 3.10.28), and the rocket exhaust could cause a substantial depletion of the atmospheric ozone layer, with a concomitant increase in ultraviolet radiation reaching the earth's surface (DOE, 1979a: 3.10.21). (Note that only a maximum of sixty flights is currently possible in any one year; the actual number is likely to be less.) Reprocessing of the spent fuel to remove 99.5 percent of the uranium would reduce the required number of flights to thirty per year. More extensive partitioning to remove short-lived fission products and plutonium, leaving only long-lived actinides, could further reduce the required shuttle launch rate. (However, as described later in this chapter, partitioning technology is not sufficiently advanced to allow this.)

Calculations by the Department of Energy suggest that the cost of space disposal of spent fuel would be prohibitive, increasing the cost of electricity to the consumer as much as 50 percent due to space shuttle costs alone (DOE, 1979a: 3.10.19). According to DOE, disposal of reprocessed waste would only increase the cost of electricity by 2 percent. At the present time, however, these estimates must be considered speculative. The total cost of waste disposal to the consumer would, of course, be greater, depending upon the cost of the waste-partitioning and packaging technology. From a theoretical point of view, space disposal appears feasible, but on the practical side, there remain a number of technical problems that may well prove insoluble. The space program has not been without its spectacular failures, sometimes as many as two within one week (*New Scientist*, 1977: 5). A failure during the rocket boost of a waste shipment could end in uncontrolled reentry and burnup of the cargo, resulting in the injection of massive quantities of radioactive material into the upper atmosphere. This is hardly a desirable method of disposal. In the absence of burnup, a more localized, but intense, contamination of land or water could occur. Once in orbit, failures in the guidance or propulsion systems of the kickstage could leave the waste in an earth orbit, that, as was the case with Skylab, would eventually decay, or in an abnormal solar orbit, which might, at some time, result in a reencounter between earth and the waste. In both instances, the unprotected waste canister would reenter the atmosphere. As DOE has pointed out, "Reliabilities of the booster vehicle, upper stages, and safety systems envisioned for the space disposal mission have not yet been determined by NASA. Reliabilities of these systems are expected to be extremely high" (DOE, 1979a: 3. 10.15). Even so, without 100 percent reliability, failures will occur. No one can guarantee 100 percent reliability. At the very least, one of the requirements of a safe space disposal system would be the development of a waste container capable of surviving reentry and ground impact.

More practical problems remain to be dealt with prior to development of a space disposal program. The success of the space shuttle program has yet to be demonstrated. Reprocessing and partitioning for such a program pose special requirements and may substantially increase worker radiation exposure over more down to earth waste management options. Handling and transportation of space disposal waste containers may be problematic. A space disposal system will be much more expensive than a terrestrial disposal system, which in any case would have to be developed for the high volume, residual high level waste that could not be sent into space.

Despite these and other drawbacks, space disposal does have its supporters. NASA is interested for obvious reasons, and certain prominent individuals, such as former Energy Secretary James Schlesinger and Senator (R-NM) and former astronaut Harrison Schmitt, are reported to favor strongly the idea. The concept is presently under study by NASA and DOE, and a decision on whether or not to proceed with space disposal will be made in 1981. However, given the outstanding questions involving safety, cost, and undeveloped technology, we do not believe that space disposal can be considered a serious radioactive waste management option for the foreseeable future (see ERDA, 1976; APS, 1977; OSTP, 1978b; DOE, 1979a).

Ice Disposal

Disposal of hot radioactive wastes in the continental ice sheets of Greenland or Antarctica has been offered as an "international" solution to the problem. First proposed by Bernard Philberth, who received a German patent on the idea in 1958, ice disposal was revived by a group of scientists in 1973 (Zeller, Saunders, and Angino, 1973). The concept has several inherently attractive features: ice disposal would offer geographical isolation, could offer long-term isolation of the wastes from the environment, and would not require the development of any "exotic" technology in order to implement. Ice itself has several advantages as a disposal medium: like salt, its fractures are self-healing through recrystallization and plastic flow, it is impermeable to water, and its low temperatures make ice a good heat sink for hot radioactive wastes. However, political uncertainties plus a number of unresolved technical questions, discussed below, tend to make ice sheet disposal a dubious, if not entirely unacceptable, proposition. Three ice disposal concepts have been proposed—meltdown, anchored emplacement, and surface storage. These are shown in Figure 3–12.

In the meltdown concept, the hot wastes would be encapsulated in suitably shaped metal canisters, placed in holes drilled into the ice and spaced about one mile apart, and allowed to melt through the ice at the rate of three to six feet per day. A canister would reach bedrock in a typical period of five to ten years. Because of the level of heat output required for meltdown, this concept would be suitable only for solidified high level waste or spent fuel aged less than two years.

Anchored emplacement is similar to the meltdown concept except that it would allow retrievability of the waste canisters for a certain period of time. Canisters would be placed in holes 150 to 300 feet deep, anchored to the surface by 1,000 foot long cables. The canisters would melt their way into the ice until restrained by the anchor cables. Retrievability would be possible for 200 to 400 years. Eventually, new snow and ice would accumulate on the surface, covering the canister anchors. The system would reach bedrock in about 30,000 years, during which time the canister would also be subject to ice flow patterns.

Surface storage would also allow retrievability for several hundred years. The waste canisters would be placed in self-supported surface

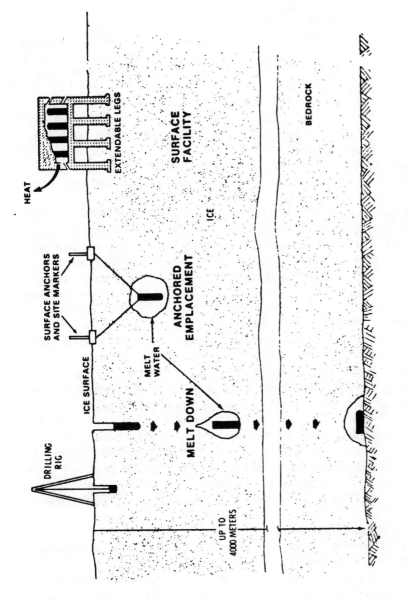

Figure 3–12. Ice Disposal Concepts.

Source: DOE (1979a:3.7.4).

storage units cooled by natural drafts. After the legs of the unit were covered with snow and ice, the unit would act as a heat source and melt down through the ice.

Uncertainties regarding the behavior of ice sheets over the long term raise serious questions about the feasibility of ice disposal. In particular, it appears that major portions of the Antarctic ice sheet are subject to periodic surges, on a scale of several tens of thousands of years, during which large volumes of ice are disgorged into the ocean (DOE, 1979a: 3.7.14). There is strong evidence to indicate the presence of water at the interface between the ice and bedrock, and it is believed that the water may act as a lubricant, thus facilitating ice surges. The presence of hot wastes at this interface could increase the volume of water present, thereby possibly precipitating surgelike behavior. In any event, an ice surge could cause ejection of the waste canisters into the ocean.

Several other uncertainties also exist. Evidence has been obtained that suggests the existence of lakes underlying the ice that may play a part in ice surges. Accidental placement of waste canisters into such bodies of water would effectively defeat the very point of ice disposal. It is also possible that radical changes in global climate, such as might be caused by the so-called "greenhouse effect," could cause widespread melting of the ice sheets. Thus, permanent isolation of the wastes by ice disposal would be by no means assured.

Another problem would involve permissible heat loading. Emplacement of too many canisters in too small an area could raise excessively the ambient temperature of the ice sheet, thereby causing widespread melting. However, wide spacing of canisters would require use of very large areas. For example, if heat load were to be limited to 1 percent of the geothermal heat flux density, only one 5.4 kilowatt waste canister could be emplaced in a ten-square kilometer area. In other words, the total disposal area required for all the spent fuel produced through the year 2000 could be several hundred thousand square kilometers (DOE, 1979a: 3.7.8; see also Zeller, Saunders, and Angino, 1973; ERDA, 1976; APS, 1977; DOE 1979a.)

Partitioning and Transmutation

Partitioning and transmutation involve, first, the chemical separation of some selected radioactive species from a mix and, second, the transmutation of that element—generally a long-lived radionuclide—into another—short lived or nonradioactive—by means of simple neutron capture or neutron-induced fission. The two processes, taken together, do not constitute a disposal technology per se, but rather are approaches to reducing the hazards of radioactive wastes by elimination of the long-lived actinides and fission products. Waste management specialists feel confident that isolation of radioactive wastes can, at least in principle, be guaranteed over the thousand or so years necessary for the shorter lived fission products to decay to relatively innocuous levels. They reason that, if some method can be found to convert the long-lived species to shorter lived or stable atoms, the problem of ensuring extreme long-term isolation will essentially disappear.

Theoretically, partitioning and transmutation could reduce the long-term hazard of the waste by a factor of one hundred. However, the technical requirements are severe. Partitioning requires reprocessing and recycle of all uranium and plutonium in spent fuel as well as the separated long-lived radioisotopes, plus the disposal of the remaining short-lived wastes. To date, partitioning has not been demonstrated on a commercial scale. In fact, separation processes, including Purex reprocessing, leave several percent or more of most elements of interest in the waste stream. A commercial process would have to demonstrate a much higher recovery percentage in order to be considered suitable for waste management purposes. Some other problems arise with partitioning, including increased production of secondary wastes, increased transportation costs and requirements, increased costs of waste management, and the increased potential for worker radiation exposure.

A successful transmutation technology must meet four criteria in order to be considered theoretically feasible and practical:

1. It must consume much less energy than is produced by the creation of the radioactive wastes;
2. It must remove more long-term, high risk radioactive waste than it creates;
3. The rate of transmutation must be greater than that of natural decay; and
4. It must be capable of eliminating a large fraction of the long-lived constituents of the waste.

In addition, a sufficient number of transmutation devices must be available for the process to be practical on a large scale (ERDA, 1976: 27.1).

Four types of transmutation devices have been proposed—charged particle accelerators, nuclear explosive devices, fusion reactors, and fission reactors. Charged particle accelerators have been deemed impractical because of unfavorably large power consumption. For obvious reasons, nuclear explosive devices are considered an impractical approach to transmutation, although in theory they would work. However, a large number of devices would be required. For example, it has been found that the fissioning of all the neptunium, americium, and curium produced annually by a 1,000 megawatt reactor would require three and a half 100 kiloton thermonuclear detonations (ERDA, 1976: 27.8). Transmutation in fusion reactors would be theoretically feasible, but as the American Physical Society study observed, "Transmutation by fusion reactors is hardly to be relied upon in the absence of fusion reactors; if they did exist, we would not be concerned with long-term incentives for actinide transmutation" (APS, 1977: S114).

Transmutation in fission reactors is generally considered to be feasible on a theoretical basis; however, the difficult technical requirements of partitioning on a large scale, including fabrication of waste into fuel elements suitable for insertion into fission reactors, and of overcoming low neutron fluxes in existing light water reactors suggest that development of a viable transmutation technology will require decades. In addition, studies have generally shown that transmutation in light water reactors of fission products such as strontium–90 and cesium–137 would not be appreciably faster than natural decay. Transmutation into innocuous elements of the iodine-129 and technetium–99 produced annually in a light water reactor would require some 50 to 500 years in a light water reactor system (ERDA, 1976: 27.5). Fissioning of the annual production of actinides by one reactor would require from ten to thirty years (ERDA, 1976: 27.10).

By mid-1979, the Department of Energy was concluding a three year study into the feasibility of and incentives for partioning and transmutation. It has been estimated that development of a partitioning and transmutation program would require at least thirty years (DOE, 1979a: 3.9.23). Also, it is not clear that, compared to the direct disposal of spent reactor fuel, partitioning and transmutation offer a superior waste management technology. Given the uncertain environmental, social, and economic costs of developing this technology on a large scale, it appears unlikely that commercialization of partitioning and transmutation will ever be achieved (see ERDA, 1976; APS, 1977; OSTP, 1978b; DOE, 1979a).

THE HAZARDS AND RISKS OF RADIOACTIVE WASTE MANAGEMENT

How hazardous are radioactive wastes, and what risks do they pose to human populations and the environment in general? These questions are of major concern to a discussion of radioactive waste management, for answering them may suggest both the magnitude of the hazard and the necessary requirements for minimizing the hazards. Hence, this section discusses the toxicity of radioactive wastes and the hazards and risks of radioactive waste management. Left unanswered, however, is the question of what constitutes an acceptable risk from radioactive waste management, which can only be answered through societal consensus.

Toxicity of Radioactive Waste

A common measure of the toxicity of radioactive material is the "ingestion hazard index," which is that volume of water required to dilute a given quantity of radioactive material to public drinking water standards based on federal maximum permissible concentrations for the relevant isotopes. For example, it would require about fifteen trillion cubic meters[2] of water—roughly the volume of Lake Superior—to dilute the annual spent fuel discharge from a 1,000 MW reactor (thirty metric tons or sixty-six thousand pounds) to safe levels one year after the fuel has been removed from the reactor. Even after one million years, more than twenty-two million cubic meters of water—enough to cover an area of twenty-eight square miles to a depth of one foot—would be required for dilution of the fuel to safe levels (Figure 3-13).

The ingestion hazard of the spent fuel may also be compared to that of naturally occurring uranium ore. By this standard, the toxicity of spent fuel one year after discharge is almost six million times that of an equal weight of typical uranium ore containing 0.2 percent uranium oxide. One million years later, the fuel is still more than 200 times as toxic as the ore (Figure 3-14).

The total ingestion hazard index of radioactive wastes currently in storage is quite large. For example, dilution of the strontium-90 and cesium-137 contained in all of the high level defense wastes currently in storage at Hanford, Idaho Falls, and Savannah River to maximum permissible concentrations would require roughly 950 trillion cubic meters of water, more than forty times the volume of all of the Great Lakes. Table 3-1 shows the partial and total ingestion hazard indexes of the defense wastes as a function of time. Dilution of the strontium-90 and cesium-137 contained in the 5,700 metric tons of spent reactor fuel estimated to be in storage at the end of 1979 to permissible levels would require about 2,800 trillion cubic meters of water, more than twenty times the volume of water in the

2. 1 cubic meter = 35.3 cubic feet.

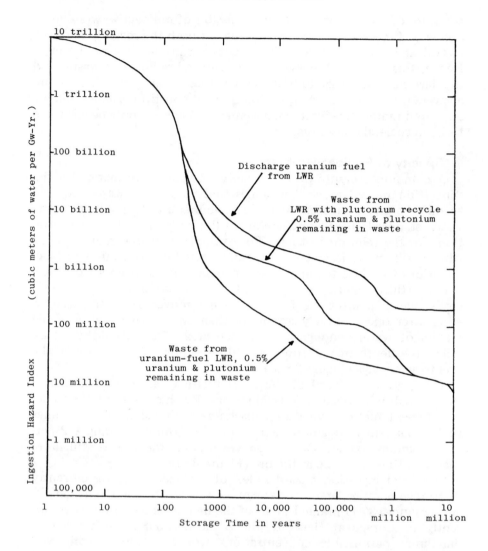

Figure 3–13. Ingestion Hazard Index of High Level Wastes and Spent Fuel from a Light Water Reactor. The spent fuel discharge from one Gw Yr. *(gigawatt year = 1,000 megawatt year)* is approximately 30 metric tons.

Source: Redrawn by author from APS (1977:S110, ref. 41).

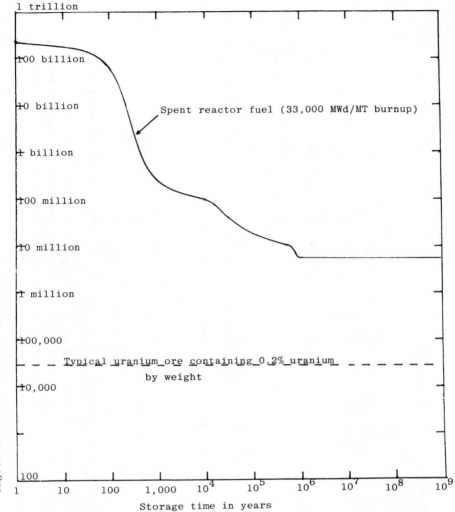

Figure 3—14. Ingestion Hazard Index per Metric Ton of Spent Fuel Discharged from a Light Water Reactor Compared to 1 Metric Ton of a Typical Uranium Ore Containing 0.2 percent Uranium by Weight.

Source: Redrawn by author from JPL (1977: A—15).

Table 3–1.   Ingestion Hazard Index of High Level Defense Wastes in Storage as of 1976[a] (cubic meters of water).[b]

Nuclide[c]	Initial Inventory (curies)	1976 (0 yr)	10 yr	50 yr	100 yr	500 yr
Sr-90	280 Mci[d]	9.4×10^{14}[e]	7.3×10^{14}	2.6×10^{14}	7.8×10^{13}	3.9×10^{9}
Cs-137	310 Mci	1.6×10^{13}	1.2×10^{13}	5.0×10^{12}	1.5×10^{12}	1.5×10^{8}
Pu-239	30,000	6.0×10^{9}	6.0×10^{9}	6.0×10^{9}	6.0×10^{9}	6.0×10^{9}
I-129	105	1.8×10^{9}	1.8×10^{9}	1.8×10^{9}	1.8×10^{9}	1.8×10^{9}
Total		9.6×10^{14}	7.4×10^{14}	2.6×10^{14}	8.0×10^{13}	1.2×10^{10}

Table 3–1. continued

Nuclide[c]	1,000 yr	10,000 yr	100,000 yr	1,000,000 yr	10,000,000 yr
Sr-90	16,700	—	—	—	—
Cs-137	930	—	—	—	—
Pu-239	5.8×10^{9}	4.5×10^{9}	3.5×10^{8}	—	—
I-129	1.8×10^{9}	1.8×10^{9}	1.8×10^{9}	1.7×10^{9}	1.2×10^{9}
Total	8.9×10^{9}	6.8×10^{9}	2.1×10^{9}	1.7×10^{9}	1.2×10^{9}

[a] Sr-90 and Cs-137 inventories from Krugmann and von Hippel (1977:884); other numbers from NAS (1978:214); Hanford inventory scaled up to include INEL and Savannah River inventories.

[b] One cubic meter = 35.3 cubic feet; see Table 3–3 for volumes of bodies of water.

[c] Includes only nuclides that dominate ingestion hazard index during any period; others contribute small fraction to hazard. MPC_w: Sr-90, 3×10^{-7} ci/cu.m.; Cs-137, 2×10^{-5} ci/cu.m.; Pu-239, 5×10^{-6} ci/cu.m.; I-129, 6×10^{-8} ci/cu.m.

[d] Mci = megacuries = millions of curies.

[e] 9.4×10^{14} = 940 trillion.

world's freshwater lakes. Table 3—2 shows the partial and total ingestion hazard indexes of the current spent fuel inventory as a function of time. Table 3—3 lists the volumes of water present on earth and in several lakes and rivers.

The validity of these comparisons as a measure of toxicity is limited in that they apply only to drinking water and cannot account for concentration mechanisms in the food chain. And of course, no one would consider dilution of large quantities of radioactivity in this way. Such comparisons are also defective from another point of view; they say nothing about the risks of radioactive waste management. These comparisons are often used to minimize the problem of radioactive wastes by those who note that other, nonradioactive substances have greater toxicity. Similarly, it is often suggested that an acceptable toxicity level for radioactive waste is one that is equivalent to the ore from which it came. Arguments such as these are specious and misleading. Toxic materials, if handled properly, can pose minimal health risks to the public. On the other hand, even materials with relatively low toxicity, if handled improperly, can be quite hazardous (see the discussion on mill tailings in Chapter 4).

Risks of Radioactive Waste Storage and Disposal

Hazard indexes are not very meaningful if one wishes to evaluate the relative hazards and risks of different waste management technologies. Such an evaluation requires a more sophisticated technique—called "risk analysis"—that assesses how and with what probability a given technology might fail, the consequences of failure, and the risks to exposed individuals.[3] It is extremely difficult to perform a risk analysis of a complex technological system for which little or no operating experience has been accumulated, as is the case with radioactive waste disposal. At best, it is possible to postulate a finite number of events (failures), set upper and lower bounds on their probability of occurrence, determine the consequences, and decide if the risks are acceptable now and will continue to be acceptable in the future. The uncertainties that are associated with such steps may be very large, indeed, large enough in some cases to make predictions meaningless. In what follows, we discuss the modes of failure of a deep geological radioactive waste repository.

---

3. "Probability" is used here to mean the annual chance that a given technology will fail; "consequence" is equal to the number of deaths resulting from failure of a technology; and "risk"—the product of probability and consequence—is the annual chance of dying as a result of a postulated failure.

Table 3-2.   Ingestion Hazard Index of Spent Fuel in Temporary Storage as of December 1979[a] (in cubic meters of water).[b]

Nuclide[c]	Initial Inventory (curies)	1979 (0 yr)	10 yr	50 yr	100 yr	500 yr
Sr-90	820 Mci[d]	2.7×10^{15} [e]	2.1×10^{15}	7.9×10^{14}	2.3×10^{14}	1.2×10^{10}
Cs-137	1,100 Mci	5.5×10^{13}	4.4×10^{13}	1.7×10^{13}	5.5×10^{12}	5.3×10^{8}
Tc-99	85,500	2.8×10^{8}	2.8×10^{8}	2.8×10^{8}	2.8×10^{8}	2.8×10^{8}
Pu-239	1,840,000	3.7×10^{11}	3.7×10^{11}	3.7×10^{11}	3.7×10^{11}	3.7×10^{11}
I-129	250	4.2×10^{9}	4.2×10^{9}	4.2×10^{9}	4.2×10^{9}	4.2×10^{9}
U-238	1,700	4.2×10^{7}	4.2×10^{7}	4.2×10^{7}	4.2×10^{7}	4.2×10^{7}
Ra-226[f]	—	—	—	—	—	—
Total	—	2.8×10^{15}	2.1×10^{15}	8.1×10^{14}	2.4×10^{14}	3.9×10^{11}

Table 3-2. continued

Nuclide^c	1,000 yr	10,000 yr	100,000 yr	1,000,000 yr	10,000,000 yr
Sr-90	5.2×10^4	—	—	—	—
Cs-137	5,100	—	—	—	—
Tc-99	2.8×10^8	2.7×10^8	2.0×10^8	1.0×10^8	—
Pu-239	3.6×10^{11}	2.8×10^{11}	2.2×10^{10}	—	—
I-129	4.2×10^9	4.2×10^9	4.2×10^9	4.0×10^9	2.8×10^9
U-238	4.2×10^7	4.2×10^7	4.2×10^7	4.2×10^7	4.2×10^7
Ra-226^f	—	1.9×10^9	1.9×10^{11}	8.9×10^{10}	8.9×10^{10}
Total	3.6×10^{11}	2.9×10^{11}	2.2×10^{11}	9.3×10^{10}	9.2×10^{10}

^{a} 5,700 metric tons, from IRG (1979:D-28). Assumes average age of approximately three years.

^{b} One cubic meter = 35.3 cubic feet; see Table 3-3 for volumes of bodies of water.

^{c} Includes only nuclides that dominate the ingestion hazard index during any period; others contribute small fraction to hazard; MPC$_W$ (curies per cubic meter): Sr-90, 3×10^{-7}; Cs-137, 2×10^{-5}; Tc-99, 3×10^{-4}; Pu-239, 5×10^{-6}; I-129, 6×10^{-8}; U-238, 4×10^{-5}; Ra-226, 3×10^{-8}.

^{d} MCi = megacuries = million of curies.

^{e} 2.7×10^{15} = 2,700 trillion.

^{f} Ra-226 is present in very small quantity in freshly discharged reactor fuel; it builds up as a consequence of decay of heavier actinides.

Table 3–3.   Volumes of Water on Earth and in Several Lakes and Rivers.

Volumes of Water on Earth (cubic meters):

Oceans	1.4×10^{18}
Rivers (average capacity)	1.7×10^{12}
Freshwater lakes	1.2×10^{14}
Saline lakes and inland seas	1.0×10^{14}
Soil moisture and vadose water	1.5×10^{14}
Groundwater (to 1,000 meter depth)	7.0×10^{15}
Ice caps and glaciers	2.6×10^{16}
Annual river runoff	3.0×10^{13}
Total	1.43×10^{18}

Source: Bredehoeft et al., (1978:2).

Volumes of the Great Lakes (cubic meters):

Superior	1.22×10^{13}
Michigan	4.92×10^{12}
Huron	3.54×10^{12}
Ontario	1.64×10^{12}
Erie	4.83×10^{11}

Source: National Geographic Society (1973).

Annual Flow of Some Large Rivers (cubic meters):

Hudson	1.0×10^{9}
Columbia	2.3×10^{11}
Mississippi-Missouri	5.5×10^{11}
Amazon	3.6×10^{12}

Source: Lapedes (1974:473).

Repository Failure and Consequences.   We here assume that high level waste and spent reactor fuel is buried in deep geological formations and successfully isolated for the period of time during which the wastes are monitored (about one hundred years). How might the wastes be released thereafter?[4] There are three plausible modes of repository failure that could lead to release of wastes: (1) catastrophic (sudden), (2) noncatastrophic (slow), and (3) human-induced. What would be the consequences of repository failure? Sud-

4. Most of this section is based upon the EPA (1978b) study prepared by Arthur D. Little, Inc. (ADL), Cambridge, Massachusetts, which attempts to evaluate the probability of repository failure and the consequences and risks associated with such failure.

den dispersion of the wastes into the air (by meteorite, for example) could have widespread and serious consequences distributed over a relatively short time. Slow release of wastes, on the other hand, could have small annual consequences, but continuing for many thousands of years. Risk is highest for high probability, high consequence events and lowest for low probability, low consequence events.

Catastrophic release could come about because of meteorite impact on or volcanic or seismic disruption of a repository, resulting in significant short-term consequences.[5] The probability of such an event is quite small (about one in one billion per year), and so, on an annual basis, the risk of injury as a result of catastrophic failure is also quite small.

Noncatastrophic release could occur as a result of major groundwater intrusion into the repository, followed by leaching of the waste and eventual transport into the biosphere. Over the long term this is the most likely failure mode. Indeed, it is virtually certain to occur. Groundwater intrusion could, for example, be caused by activation of a previously inactive, undetected fault or by climatic changes in the repository region. Unless the failure occurs during the first few hundred years, the annual consequences and, hence, individual risk are likely to be quite small. Over the lifetime of the repository, however, the cumulative consequences could be very large, even though the annual individual risk might be quite small (for example, one death per year would result in one hundred thousand deaths over 100,000 years).

Human-induced release is the least quantifiable failure mode because predictions about the future behavior of individuals and societies cannot be made with any accuracy. Repository failure might be caused by sabotage, war, or a search for minerals. Sabotage would require a major commitment of skill and resources and is thought to be of low probability. In the event of nuclear war, more appropriate targets and more pressing concerns would no doubt present themselves. A search for mineral resources appears to be the most likely cause of repository disruption, particularly if the facility is located in salt.[6]

Salt is frequently associated with oil, gas, potash, and other minerals and is itself valuable. Intrusion into the repository is a high probability, low consequence event, posing little risk to a large population.

5. Short term is zero to one hundred years; long term is one thousand to one million years.

6. The ADL study estimates a probability of 0.01 per year that petroleum exploration activities will lead to breaching of the repository. The chances of hitting an individual canister are much smaller.

Breaching the facility could, however, open the way to groundwater intrusion and noncatastrophic failure.

Another high probability event is repository disruption by petroleum extraction activities, such as hydraulic fracturing,[7] which might allow groundwater to enter the facility. Because petroleum extraction and exploration is not likely to be restricted outside of a small buffer zone around the repository, such disruption could occur early in the repository lifetime and might lead to moderately severe consequences and risk.

Finally, another important resource is water. Extensive irrigation or extraction of water in one part of a geological water basin could seriously alter the regional groundwater configuration, possibly disrupting a waste repository located in another part of the same basin. The consequences and risk could be significant, particularly if radionuclides were drawn into the water being used.

Of course, one could imagine a hundred or more other ways in which repository failure could come about. Unfortunately there is no way to take all failure modes into account. Many high probability, low consequence and common mode failures not now identified or incorrectly assigned low probabilities could nonetheless pose some substantial risk.

Risks of Waste Management Technology Failure. Risk analyses of complex, untested technologies generally require that certain simplifying assumptions be made, but this practice tends to yield quantitative risk assessments with very large ranges of uncertainty. Thus, an upper bound to a calculated risk may be some ten to one thousand times the average value of the risk, with a lower bound comparably smaller. It is more appropriate, therefore, to consider a risk analysis as an order of magnitude approximation rather than an absolute quantification of risk. Even so, assessment of the risks of waste management technologies can be useful. We discuss three such analyses below. The first, performed for the Environmental Protection Agency, will provide a basis for environmental radiation release standards to be promulgated by the agency. The second has been widely used as "proof" of the safety of geologic isolation, even though it omits several critical considerations. The third is a simple, hypothetical analysis, performed for this book, that assesses the consequences of several types of radioactive waste facility failures.

---

7. Hydraulic fracturing involves injection of water under high pressure into oil-producing formations in order to extract petroleum unrecoverable by ordinary extraction techniques.

1. The Arthur D. Little study (EPA, 1978b) has "calculated" that over the first 10,000 years of repository operation, some deaths are virtually certain to occur as a result of radioactive waste release.[8] Oddly enough, the greatest risk comes not from a catastrophic event but rather from the effects of groundwater leaching and transport of technetium–99, a highly mobile, long-lived fission product, simply because this radionuclide remains radioactive long enough to reach the accessible environment through groundwater pathways. Catastrophic events are deemed to be of such low probability that the annual risk per individual is quite small. The ADL study concludes that the probability of significant human disruption is very small, although not as small as the probability of catastrophic disruption.

2. A quantitative analysis of the risks of noncatastrophic repository failure has been prepared by nuclear proponent Bernard Cohen (1977a & b) of the University of Pittsburgh. He assumes that the solidified wastes produced by the reprocessing of spent fuel from one year of all-nuclear domestic electrical generation (equal to 400 gigawatts or about 400 large reactors producing some 28,000 cubic feet of solidified waste each year) are successfully isolated in bedded salt at a depth of 600 meters for 500 years.[9] He then assumes that groundwater enters the facility and begins to leach away the waste and that an atom of the waste has the same chance of escaping into the environment and being ingested as an atom of radium–226 located in the top 600 meters of the United States. On this basis, Cohen finds that in the first million years after disposal, 0.4 deaths would occur from waste-induced cancer, with an additional 4 deaths in the following one hundred million years.

Are these numbers reasonable? If correct, Cohen's analysis appears to indicate that, if nothing goes wrong, geologic isolation is virtually risk-free. But in fact, some of his assumptions are questionable—perhaps too simplistic—and greatly affect the magnitude of the risk. For example, Cohen treats the waste as if it were composed entirely of radium–226, a radionuclide whose movement in groundwater may be significantly retarded by chemical reactions with surrounding rock. Other radionuclides, such as technetium–99, can move much more rapidly than radium. Moreover, several instances have been observed in which normally retarded isotopes, such as plutonium–239, have migrated great distances for unexpected reasons. Cohen

8. Only relative risks have been determined in this study. Apparently, the numbers calculated are not "hard" enough to allow determination of absolute risks (as was done, for example, in the now discredited Reactor Safety Study).

9. This quantity of 500 year old waste, if uniformly distributed over the surface of the United States, could cause 10 million fatal cancers.

gives no consideration to inactive or undetected faults that may intersect the repository (never excluded entirely even in so-called earthquake-free regions),[10] climatic changes over long periods of time that could cause shifts in the local groundwater regime and allow water to enter the repository, and chemical and thermal problems that could facilitate migration by burying hot radioactive wastes in cold rock. Moreover, in the time period of interest, major climatic change could cause erosion or removal of surface materials to depths that would render such risk estimates meaningless. In criticizing Cohen's approach to this problem, one study pointed out that:

> Radium is not emplaced in a formation by deliberately creating an artificial opening and compromising the integrity of the formation. The integrity of uranium deposits from which radium comes is the result of natural processes which have come to equilibrium over long periods of geologic time. Radium is in chemical and thermal equilibrium with its ore. By contrast, emplaced waste would not be in equilibrium with the host rock of a depository, and these physical and chemical disequilibria may be responsible for . . . important failure processes.[11] (CERCDC, 1978a: 198)

In short, any process that disrupts a repository so as to allow intrusion of water is likely to result in a release of radionuclides with increased risk to the exposed population.

The most important omissions from Cohen's analysis, however, are not natural, but human-induced failures that could come about through carelessness, irresponsibility, intentional ill-will, or simply by accident and result in greatly increased hazards. And in focusing only on deaths as an indicator of damage, Cohen ignores the societal disruptions that could result from an unexpected release of radioactive materials. The history of waste management in the United States clearly demonstrates the importance of human-induced failures and emphasizes the need to consider the human element in any risk analysis.

3. What are plausible consequences of repository failure? In order to establish a rough idea of these, three hypothetical scenarios, not the worst possible, were analyzed for the purposes of this report (see Appendix B for details). The analyses were based upon Cohen's dose calculations and assume dose-risk relationships, several times greater than Cohen's, used by the Environmental Protection Agency

10. This is apparently the case at the Waste Isolation Pilot Plant site in New Mexico where construction of a waste repository is planned; see Chapter 5.

11. This suggests that if the wastes can be made to come into equilibrium with the host rock, security will be improved. This idea is under serious study, as discussed earlier in this chapter.

and the National Academy of Sciences. Simplistic assumptions made in each analysis may overestimate or underestimate effects. Thus, the uncertainties may be large. This is, unfortunately, unavoidable. In each case, a quantity of radioactive waste was assumed to enter a river with an annual flow of 500 million cubic meters (about half that of the Hudson River), which supplies drinking water to a population of 500,000. Exposures were assumed to occur over a one year period,[12] and deaths from cancer were assumed to occur during the thirty years following a fifteen year latency period. The three scenarios and results follow:

Case 1.   We base the first calculation on experience at the Hanford Reservation (see Chapter 4). We assume that periodic leaks from aging storage tanks allow approximately 400,000 gallons of high level liquid waste containing 500,000 curies of radioactivity to escape into the ground. Of this quantity, approximately 50 percent is strontium–90. After a one hundred year delay, 10 percent of the strontium–90 in the waste reaches the river.

Results: Between 1,500 and 3,000 fatalities, or 50 to 100 per year, were found to occur as a result of leukemia and bone cancer. This is an increase of 6.25 to 12.5 percent in the cancer death rate (of 800 per year per 500,000 population).

Case 2.   One hundred years after the sealing of a salt repository containing 100,000 metric tons of spent fuel, an undetected fault becomes active and allows groundwater to enter the repository and carry away radioactive waste. During one year, about 7.5 metric tons[13] are leached away. One hundred years later, 10 percent of the strontium–90 in the leached waste enters the river.

Results: Six hundred and thirty-eight to 1,275 fatalities, or 21 to 42 per year, were found to occur as a result of leukemia and bone cancer. This is a 2.6 to 5.2 percent increase in cancer fatalities.

Case 3.   Ten thousand years after the sealing of a salt repository containing 100,000 metric tons of spent fuel, an undetected fault becomes active and allows groundwater to enter the repository and carry away radioactive waste.

---

12. Possibly unreasonable; contamination might be detected much earlier.

13. The leach rate is one-half that assumed in the Arthur D. Little study referred to previously (0.015 percent of the repository contents per year).

During one year, about 7.5 metric tons are leached away. One thousand years later, 100 percent of the leached technetium-99 enters the river during one year.

Results: There were 0.11 fatalities per year or 23,100 during the first half-life of the isotope. This would not be statistically detectable in the general cancer death rate, however.

To give some idea of the worst possible consequences of a waste facility failure, we consider the following scenario: 250,000 curies of strontium-90 leak into the river over a period of one year and enter into the drinking water supply of a city of 500,000. What are the long-term effects of a one day exposure to this contaminated water? We found that approximately 500 to 1,000 fatalities, or 17 to 34 per year, would occur as a result of leukemia and bone cancer, even though the exposure was limited to one day. This is an increase of 2 to 4 percent in the annual cancer death rate.

Are these numbers reasonable? Possibly, but they cannot be considered more than order of magnitude estimates. These scenarios are intended to demonstrate that serious consequences may arise from circumstances that are by no means the worst imaginable. Our conclusions about the relative risks of waste management technologies, based on these examples and other considerations, are:

1. *Temporary storage poses the greatest risk:* Storage facilities are likely to fail during the period of greatest hazard—the few hundred years before the large quantities of fission product wastes have decayed away—and pose substantial risk.

2. *Early repository failure could pose some risk:* A significant waste release from a repository during the first several hundred years could pose significant risk to a locally exposed population.

3. *Late repository failure probably poses little or no risk:* Slow release of waste would probably not cause a death rate statistically detectable among the general cancer death rate. This does not necessarily imply an acceptable risk, however.

Thus, on average, geological disposal appears to be a satisfactory means of isolating radioactive wastes, but certain types of failures could render a repository dangerous to an exposed population.

These analyses are marked by one significant omission: no account has been taken of nonfatal disease, genetic mutation and birth defects, denial of water supplies, and disruption of everyday life in the affected region. Nor do these analyses take into account the psychological impact upon the exposed population. Would they, like some of the thousands of citizens of Seveso, Italy, exposed in July 1976 to toxic dioxin gas, have to live the rest of their lives dreading the discovery of a latent cancer? Indeed, to the physical consequences of repository failure must be added the psychological consequences. In none of these cases can the damage be calculated or the related probability assessed. Thus, it cannot be proven that a given disposal technology is virtually "risk-free."

The theoretical basis for safe radioactive waste management does exist, although, as we have shown in this chapter, it is by no means a complete basis. Some technologies—in particular, those based on geologic isolation—appear quite promising. Indeed, it is the geologic technologies which form the centerpiece of the current federal waste management program. In Chapter 5 we describe how and where geologic isolation is being researched and developed for eventual implementation. In the following chapter, however, we adopt an historical perspective and discuss the origins and less than favorable history of radioactive waste management in the United States.

※   *Chapter 4*

The History of Radioactive Waste Management

To the reader unfamiliar with the history of the U.S. waste management program, it might seem odd that radioactive wastes are the cause of so much controversy in the United States and abroad. To be sure, the materials can be dangerous if improperly handled, but safe management of these wastes would not appear superficially to be an insurmountable problem. Unfortunately, what is possible in theory is not always realized in practice. To a great extent, the history of radioactive waste management supports this. In the past, the technologies necessary for safe management of these wastes were implemented poorly or not at all. The institutions responsible for waste management were generally at fault, proving unequal to the requirements of the task. Institutional actions often tended to exacerbate rather than resolve problems. In the excitement of the developing Atomic Age, there was little interest in the mundane problem of radioactive waste. The results were carelessness, mistakes, inflated claims, and unfulfilled promises on the part of the agencies in charge of the waste management program as well as repeated leaks of radioactivity into the environment.

The most conspicuous examples of mismanagement occurred under the aegis of the Atomic Energy Commission, now abolished. The commission, charged with both the promotion and the regulation of nuclear power, was uninterested in the radioactive waste problem, and repeated leaks of waste into the environment clearly evinced the low priorities assigned to the problem. Although this conflict of interest was alleviated to some degree by the breakup of the AEC—with regulation of radioactive waste disposal assigned to the Nuclear Regulatory Commission and implementation of the program to the Energy Research and Development Administration, later the Department of Energy—some of the same kinds of institutional weaknesses that occurred under the AEC are still apparent in the program. Among other things, many NRC and DOE staff members were formerly part of the AEC. Their low level of interest in the waste problem continues, in many cases, to this day.

The current managers of the radioactive waste management program are thus forced to confront a less than favorable record in attempting to implement the program. This chapter describes some of the more conspicuous failures in the areas of high level waste management, reprocessing, low level waste disposal, and mill tailings management.

HIGH LEVEL WASTE

The radioactive waste management program of the United States originated as a by-product of the wartime effort to develop the atomic bomb. Nuclear reactors, constructed at the Hanford Reservation near Richland, Washington, produced plutonium, which was then extracted from the spent fuel in on-site facilities. Reprocessing produced large volumes of radioactive waste, but under the pressure of the war effort, there existed neither the time, the money, nor the inclination to deal with the problem on anything but an interim basis. Furthermore, the technical ability to dispose of the wastes safely did not then exist, and so the wastes were neutralized and placed in temporary storage tanks.

These wastes are still in temporary storage. Because high level wastes are so radioactive, so hazardous, and hence so visible, repeated leaks of waste and the failure to develop disposal technologies have captured much attention. To be sure, future problems will not be identical to those of the past. Current handling of the wastes has improved considerably over past practices, but the problems of developing a suitable disposal technology are still formidable, are not yet solved, and may well lead to failures similar to those described here.

The Tanks of Hanford Reservation

During their operating lifetimes, the reprocessing plants at the Hanford Reservation turned out tens of thousands of pounds of weapons grade plutonium and almost fifty million gallons of highly radioactive liquid waste. The acidic waste was chemically neutralized and stored in single-walled carbon steel tanks (Figure 4-1). It was

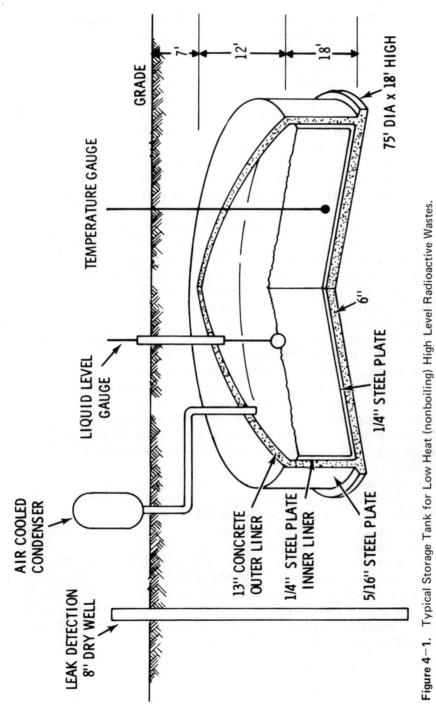

Figure 4—1. Typical Storage Tank for Low Heat (nonboiling) High Level Radioactive Wastes.

Source: AEC (1974b:II.1—73).

believed at the time that a waste disposal technology could be developed long before the failing integrity of the tanks became a problem. Problems were not long in developing, however, while the needed technology never did. In 1953, the U.S. Geological Survey observed that the tanks constituted a "potential hazard" and concluded that "their true structural life . . . [is] not known" (Gillette, 1973: 730). Despite this warning, the Atomic Energy Commission made no effort to discontinue use of the tanks. In 1957, a National Academy of Sciences panel recommended that high level waste be buried in bedded salt formations, the first of a long series of similar recommendations. The AEC disregarded this suggestion, too. In January 1959, the manager of Hanford predicted that the storage tanks would remain serviceable for "decades" and possibly as long as 500 years. Ironically, it was later discovered that a leak of 35,000 gallons, the first in a long series, had occurred some six months previous to this confident prediction (Gillette, 1973: 730). A 1968 study by the General Accounting Office, kept secret until 1970, concluded that the service life of the tanks would be only ten to twenty years (Gillette, 1973: 730). Warnings and recommendations for improved waste management were continually ignored by the AEC.

The Hanford tanks continued to spring periodic leaks, gathering much publicity in the process, and by 1973, 422,000 gallons of liquid waste containing 500,000 curies of radioactivity had seeped into the sandy soil of Hanford. One of these, involving 115,000 gallons of liquid containing 54,000 curies of activity, went undetected for fifty-one days because no one bothered to compare the readings of liquid level in the tank from one week to the next even though the readings were taken and were available to the responsible officials (Gillette, 1973: 728).

In an effort to halt the existing leaks and prevent new ones, a program of solidifying the liquid waste in the tanks was initiated at Hanford in the late 1960s (Figure 4-2). While solidification will prevent further leakage from the older tanks, it is likely to complicate the ultimate disposal of waste. Some of the waste is in the form of sludge at the bottom of the tanks, and it is not clear that this sludge can be removed. Other tanks were built with cooling coils, and the solidified waste now embeds the coils and probably cannot be extracted. Because of this, much of the waste may prove to be irretrievable, raising the question: Who is to guard the Hanford tanks for the next several thousand years?

Figure 4–2. Solidified Salt Cake Inside a Hanford Waste Tank.

Source: DOE Richland Operations Office.

The Lyons, Kansas, Debacle

Although the National Academy of Sciences recommended the burial of radioactive waste in salt formations in 1957, it was not until 1963 that the Atomic Energy Commission began Project Salt Vault, with the goal of determining whether such disposal was feasible. The Salt Vault experiments were conducted in the abandoned Carey salt mine under the town of Lyons, Kansas. The project was not overly ambitious; it consisted of the placement of a small number of waste-simulating electrically heated canisters and some canisters of real waste in holes bored into the salt. (A photo of these canisters may be seen in Figure 3–5.) It was hoped that the experiment would allow determination of the behavior of salt under conditions of high temperatures and the effects of salt upon the canisters. In fact, the salt did have some deleterious effects upon the canisters. When the heated canisters were removed from their holes, cracks penetrating halfway through the stainless steel walls were noted, and heavy corrosion was observed on the stainless steel conduits supplying power to the heaters (APS, 1977: S113). These effects were not thought to be serious or to present an obstacle to salt disposal. Still, although the AEC felt that valuable evidence bearing on the feasibility of salt burial had been gathered, the agency made no further move to develop a disposal facility.

Events, however, were moving the AEC toward development of such a facility. A 1966 National Academy of Sciences report roundly condemned waste management practices at then existing waste storage sites. Among other things, the report warned:

1. Considerations of long-range safety are in some instances subordinated to regard for economy of operation;
2. Some disposal practices are conditioned on overconfidence in the capacity of the local environment to contain vast quantities of radionuclides for indefinite periods without danger to the biosphere; and
3. "[N] one of the major sites [Hanford, Idaho Falls, and Savannah River] at which radioactive wastes are being stored or disposed of is geologically suitable for safe disposal of any manner of radioactive wastes other than very dilute, very low-level liquids" (Metzger, 1972: 153).

Rather than acting upon these conclusions, the AEC suppressed the report. Indeed, it was only under pressure from the U.S. Senate that the report was finally released in 1970.

In 1969, a serious fire took place at the Rocky Flats plutonium-processing plant near Denver, Colorado. High levels of plutonium contamination required that several damaged buildings be dismantled. These wastes were shipped to the Idaho Falls burial ground. This, however, greatly upset some of the citizens of that state, and as a result, Senator Frank Church of Idaho pressured AEC Commissioner Glenn Seaborg into promising removal of all transuranium-contaminated waste from Idaho by 1980 (Seaborg, 1971: 1582).

These two events contributed in part to the AEC's decision to develop a waste disposal facility at the Project Salt Vault site under Lyons, Kansas. By 1971, preparatory work was considered complete, and the AEC announced that permanent disposal of both transuranic and high level wastes would soon commence in the Lyons mine. The agency claimed that research had shown the method to be fully developed and safe and the site to be suitable. The wastes would be isolated from the biosphere for the requisite period of time, according to the commission's claims.

As of 1971, the AEC had spent fifteen years and some $100 million studying bedded salt disposal. The agency maintained that the radioactive waste problem was solved and that implementation of the solution could commence promptly. So confident was the AEC that Milton Shaw, director of the AEC's Division of Reactor Development, in testifying before the Joint Committee on Atomic Energy on the fiscal year 1972 budget for the Lyons project, requested the entire appropriation of $25 million for the facility rather than just that for a single year. Shaw stated that the site was "equal or superior to the others [in the country]" (Shaw, 1971: 1449).

But AEC confidence was, as things turned out, badly misplaced. The site was grossly unsuitable. The Kansas Geological Survey concluded that the Lyons site was not at all adequate for the purpose. Independent geologists concurred, for at the other end of the town of Lyons was another salt mine, and underneath the town, the galleries of the two mines came as close as 500 yards to the selected site. The operators of the second mine utilized a salt extraction process called hydraulic fracturing, in which fresh water is injected into one borehole and emerges from another as brine, which is then pumped to the surface. One of these operations unexpectedly produced no brine; over 180,000 gallons of water were pumped into a borehole and simply disappeared (Hambleton, 1972: 15). It became clear that the geology of the site was inadequately known. Moreover, the site had previously been surveyed for gas and oil. According to an AEC staff report: "In the course of drilling small holes . . . water started leaking into the mine . . ." because one of the many gas or oil boreholes in the area had been intercepted (AEC, 1971: 3). The location of many of these boreholes was not known because the state of

Kansas did not require registration of dry wells. Dr. William Hambleton of the Kansas Geological Survey described the Lyons site as "a bit like a piece of Swiss cheese" and, with respect to the AEC project, commented: "There is nothing more important than recognizing a dead horse early and burying it with as little ceremony as possible" (Hambleton, 1972: 18).

The Kansas Geological Survey recommended abandonment of the site, and the citizens of Kansas showed strong resistance to the idea of having a radioactive waste disposal site in their state. Under pressure from the Kansas congressional delegation, the AEC announced in 1973 that the Lyons site was being abandoned. Thus, after seventeen years, the search for a solution was back at square zero, clearly the result of institutional incompetence. Certainly, in no way did the AEC's actions indicate that salt disposal was a technically unsound concept, but a minimal amount of searching for old oil and gas wells would have immediately shown the unsuitability of the Lyons site for permanent waste disposal.

In the aftermath of the failure at Lyons, the AEC was hard pressed to find an alternate solution and so turned once again to the idea of interim storage, previously abandoned due to the large burden imposed by the need for long-term surveillance. The agency's answer to the high level waste disposal problem was the Retrievable Surface Storage Facility (RSSF), a temporary facility that, it was alleged, would provide safe storage of commercial and military waste for up to one hundred years. The RSSF was to be a large field of many concrete mausolea. Into each structure, one canister of high level waste would be inserted (Figure 4–3) (Carter, 1977: 664). The plan received a great deal of favorable publicity, most of it generated by the AEC and its successor agency, the Energy Research and Development Administration. Nonetheless, the Environmental Protection Agency, in reviewing the environmental impact statement on the project, gave the RSSF the lowest possible rating, expressing the well-founded fear that the facility might become a permanent, low budget substitute for a more secure method of disposal. This rejection and unfavorable public reaction caused ERDA to announce that it was dropping the RSSF and returning to the investigation of bedded salt disposal, this time in New Mexico. There the matter rests at this time.

While the history of high level waste management is notable for its carelessness, irresponsibility, false claims, and general lack of successes, other parts of the back end of the fuel cycle have fared little better, as we discuss next.

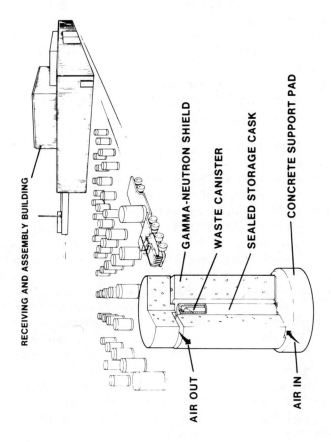

Figure 4–3. The Retrievable Surface Storage Facility (RSSF).

Source: NRC (1976b: 4–33).

REPROCESSING

In 1954, the U.S. Congress passed the Atomic Energy Act with the intention of sharing the federal government's monopoly of atomic energy with a commercial civilian power industry. Involved were all aspects of the nuclear fuel cycle except uranium enrichment. In 1956, the Atomic Energy Commission expressed its willingness to share information on the chemical reprocessing of nuclear fuels with private industry, and by 1963, construction of a commercial reprocessing plant had begun at West Valley, New York. But the thriving reprocessing industry envisioned by the AEC has never developed, for a variety of technical, economic, and political reasons. Because the reprocessing of spent fuel is so intimately tied to the production of radioactive waste, the history of the American experience with reprocessing is described here.

The Nuclear Fuel Services Plant
at West Valley, New York

The Nuclear Fuel Services (NFS) reprocessing plant at West Valley, planned as the centerpiece of an atomic industrial park sponsored by the state of New York, began operation in 1966. The plant has had a discouraging history. Despite a reprocessing capacity of 250 tons of spent fuel annually, during six years of operation only some 600 tons of spent fuel were actually reprocessed. Because the commercial nuclear power industry was quite small when the plant came on line, the AEC agreed to provide a guaranteed quantity of fuel from its Hanford plutonium reactors. Thus, 380 tons of the spent fuel reprocessed at West Valley actually came from the N-plutonium production reactor at Hanford. The remainder came from five small commercial power reactors and a number of experimental power reactors (Rochlin et al., 1978: 22).

Because the plant was forced to operate far below capacity, it was plagued by problems. Lack of suitable fuel from commercial light water reactors forced it to handle other types of fuel that tended to damage the chemistry of the reprocessing line. For example, in 1968 NFS reprocessed an experimental thorium-based fuel core from the Indian Point-1 reactor, which required a chemical process different than the one ordinarily used. The plant management hired large numbers of temporary untrained workers to perform jobs in high radiation areas. These workers would be quickly "burned out," sometimes receiving a maximum radiation dose in a matter of minutes, a practice that was perfectly legal under then existing AEC regulations. The plant was also prone to repeated off-site leaks of radioactivity.

In 1971, the plant operated at only 26 percent of capacity. The following year, the owner of the plant, Getty Oil, decided to shut it down in order to enlarge its capacity. (As constructed, the plant was incapable of serving more than about eight large reactors per year, too few to make it an economically worthwhile venture.) The cost of expansion was initially estimated to be $15 million. However, new AEC and Nuclear Regulatory Commission regulations requiring that waste solidification facilities be installed at the plant site and needed reconstruction of the plant to meet more rigid seismic requirements drove the cost up. By 1976, modification was estimated to require in excess of $600 million, and so Getty Oil announced its intention to abandon the facility, withdraw from the reprocessing business, and pursuant to the NFS agreement with the state of New York, yield ownership of some 600,000 gallons of high level liquid waste to the state. New York has been more than a little reluctant to assume this responsibility, but it is required to do so under the terms of its agreement with the old AEC.

For a long period of time, the federal government and the Department of Energy refused to consider assumption of responsibility for the waste or assistance in the costs of disposal, which are likely to be in the range of $600 million to $1 billion. (Decontamination and decommissioning of the NFS plant will probably cost another $500 million or more.) The department did, however, offer technical assistance in the resolution of the problem. More recently, DOE and New York reached a preliminary agreement over the future of the West Valley site, which included expansion and reracking of the plant's spent fuel pool and reopening of the site's low level waste burial ground in return for federal financial and technical assistance in vitrification and disposal of the high level waste at the site (*Nucleonics Week*, 1979g: 8). However, the agreement was reached in the face of intense opposition from some members of New York's congressional delegation and from many citizens. After the Three Mile Island accident, the agreement was disavowed by New York's Governor Carey, and its exact status has become somewhat confused.

What is clear, however, is that the wastes at West Valley present a continuing long-term problem. There are two tanks of high level waste at the plant: one contains 600,000 gallons of neutralized liquid, the bulk of whose radioactivity is concentrated in an irretrievable sludge at the tank bottom; the other contains 12,000 gallons of liquid from the reprocessing of thorium fuel. In July 1978, a chemist from the Woods Hole Oceanographic Institution found elevated levels of radioactivity in Lakes Erie and Ontario. Tracing the radioactivity to its source led the scientist to conclude that it came from the West Valley site—most probably from the low level waste burial ground (*Industrial Research/Development*, 1978: 28). At the end of 1978, it was announced that the steel containment saucer below the large liquid waste tank had developed a hole (Severo, 1979). While this presents no immediate problem, should the tank ever spring a leak, there would be no way to prevent the liquid from percolating into the ground. Although there is a second identical tank near the first, transferring the waste is likely to prove very difficult and may, in fact, only serve to exacerbate the problem.

Other Reprocessing Efforts

Problems have arisen with other reprocessing facilities, too. General Electric built a 300 ton per year reprocessing plant at Morris, Illinois, near Commonwealth Edison's Dresden reactors, which was to have begun operation in 1972. The plant never worked. GE decided to abandon the "tried and true" Purex process and developed a new fluoride-based reprocessing system called "Aquaflor." The Aquaflor system tended to clog up at the point where a liquid stream of uranium was converted to a solid powder (Metz, 1977: 44). GE reached the conclusion that the plant would not work for "technical reasons" and would require an additional four years and $90−130 million to redesign and rebuild. As a result, GE decided to abandon the plant, unused, and wrote off a $64 million investment in the process (*Nuclear Industry*, 1974: 8).

Allied-General Nuclear Services, a joint venture by the Allied Chemical Company and General Atomic (owned by Gulf and Royal Dutch/Shell), has built a reprocessing facility at Barnwell, South Carolina. To date, the plant has not operated, primarily because of changing NRC regulatory requirements. After the facility was completed in 1975 at a cost of $250 million (compared to the original $100 million estimate), the NRC decided that liquid plutonium and liquid reprocessing waste would have to be solidified at the reprocessing site before being shipped off site. These additional facilities were estimated to cost between $500 million and $650 million in 1976 (Tannenbaum, 1977: 1), raising the total cost of the plant to well over $750 million. At these costs, there is some question as to whether reprocessing at Barnwell would be economically justifiable, although Allied-General has said that the plant would operate if allowed to do so. Allied-General has also tried to interest the federal government in purchasing the facility and leasing it back to the company for experimental purposes. As of 1979, the government had refused to do so.

LOW LEVEL WASTE

The problem of low level waste disposal has attracted far less attention than that of high level waste. It is a serious problem, however, for the volumes of waste produced are very large. Although the contained radioactivity is dilute, the total amounts involved are quite substantial (Table 4–1). Some idea of the volumes may be gathered from a report by a National Academy of Sciences panel that found that the solid low level waste annually generated only at government-owned locations was "approximately equivalent to the volume of solid waste produced annually by a representative town in the United States with a population of 55,000. . . ." (NAS, 1976: 1).

In the past, disposal of this waste has been haphazard, at the very least. The locations of many disposal sites associated with the Manhattan Project are now no longer known, leading the Department of Energy to request public assistance in rediscovering them. In 1975, a cache of plutonium-contaminated garbage was found in the parking lot of a Los Alamos, New Mexico, motel. Between 1946 and 1969, some 59,000 containers of low level waste were dumped at sea off the Farallon Islands near San Francisco, and according to an estimate by the Environmental Protection Agency, as many as 25 percent may be leaking[1] (*Nucleonics Week* 1977c: 12). Other waste dumps off the Delaware coast and Boston harbor have also been plagued with leaks. As of now, there is no clear evidence that these dumps pose a health hazard to humans. The dumps are, however, excellent samples of poor waste management practices.

Low level solid wastes have generally been buried at one of the six commercial or five Department of Energy disposal sites (Figure 4–4). Handling of the wastes at these burial grounds has not been well controlled. Containers have often lacked labels indicating radionuclide content or activity. About 85 to 90 percent of the low level wastes contain low levels of radioactivity; the remainder may be contaminated with transuranic nuclides or may contain "hot spots." Several years ago, the official federal definition of low level wastes was changed. The low level wastes were renamed "nonhigh level wastes" and were defined to include any waste that is not high level and contains less than ten nanocuries of alpha activity per cubic foot. All nonhigh level wastes can be buried at a low level disposal site unless

---

1. Of course, eventually all of the containers will leak.

Table 4–1. Low Level Waste Inventories in the United States as of January 1, 1977.

| Site | DOE-operated Disposal Sites | | Transuranium Waste | |
	Status (O = open; C = closed)	Low Level[a] (Millions of cu. ft.)	Buried[b] (millions of cu. ft.)	Retrievable (millions of cu. ft.)
Hanford, Washington	O	6.40	5.40	0.27
Idaho Falls, Idaho (Idaho National Engineering Laboratory)	O	5.27	2.30	1.28
Los Alamos, New Mexico	O	8.55	4.10	0.06
Oak Ridge National Laboratory	O	6.42	0.20	0.05
Savannah River, South Carolina	O	9.27	1.00	0.06
Nevada Test Site (NTS)	?	0.27	<0.01	<0.01
Sandia Laboratories, Albuquerque, New Mexico	O	0.04	—	—
Others[c]	—	14.59	—	—
Totals		50.81	13.00	1.72

Table 4-1. continued

	Commercially Operated Disposal Sites		
Site and Operator	Status	Low Level	Buried TRU[d]
		(millions of cu. ft.)	(kilograms)[e]
Barnwell, South Carolina Chem-Nuclear Systems	O	3.52	0
Beatty, Nevada Nuclear Engineering Company (NECO)	O	1.97	14.3
Richland, Washington (Hanford) NECO	O	0.51	22.7
Maxey Flats, Kentucky NECO	C	4.95	69.1
Sheffield, Illinois NECO	C	2.40	13.4
West Valley, New York Nuclear Fuel Services	C	2.46	3.6
Totals		15.81	123.1

[a]Includes previously buried TRU waste.

[b]Burial of TRU waste at DOE sites ceased in 1974.

[c]Wastes contaminated with uranium buried onsite at Pantex (Amarillo, Texas), Feed Materials Production Center (Ohio), National Lead (Albany, New York), Weldon Springs, Missouri, and gaseous diffusion enrichment plants at Portsmouth, Ohio, Paducah, Kentucky, and Oak Ridge, Tennessee.

[d]Only Richland accepts commercial transuraniuc (TRU) waste.

[e]Volumes are not known.

Source: IRG (1978).

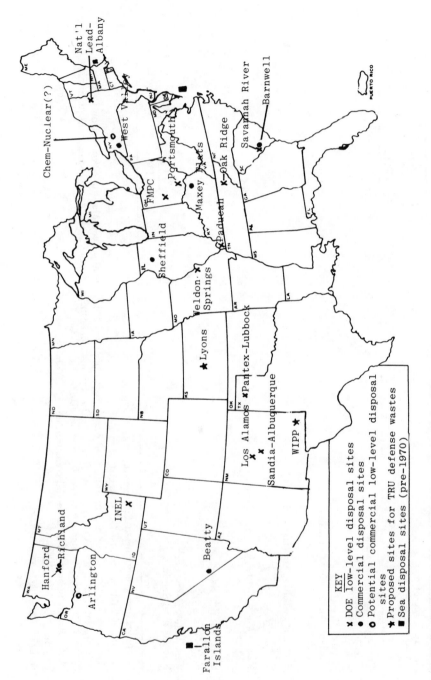
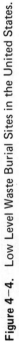

Figure 4–4. Low Level Waste Burial Sites in the United States.

Source: Prepared by author from IRG (1978: Appendix D).

specific site regulations prohibit such disposal (House of Representatives, 1976: 237). Since 1970, all wastes exceeding the ten nanocurie transuranic contamination limit have been shipped either to Idaho Falls or to the commercial site at Richland, Washington (near Hanford). At Idaho Falls the wastes are packed into barrels and stored retrievably for future disposal. The Richland wastes are buried in trenches.

Even though these low level disposal sites are subject to federal regulation (either by DOE or NRC), waste-handling practices have often left much to be desired. The disposal procedure for low level wastes at the Idaho Falls burial site was once described by a worker at the site:

> The way that they dispose of their radioactive waste is, they take heavy equipment and they go out and they dig a trench. They put the smaller particles in a pasteboard box and the rest of the stuff is dumped directly into the ground. (Metzger, 1972: 155)

The handling of wastes at the Idaho Falls site was even more haphazard than this description indicates. Newspapers once published an AEC photograph of flooded burial trenches with containers of radioactive waste floating in the mud. Some other examples of waste mismanagement are described in detail below.

Critical Mass at Hanford

At the Hanford Reservation, low level liquid wastes were dumped into shallow trenches with unsealed bottoms called "cribs" (Figure 4–5), with the expectation that ion retention processes in the soil would prevent excessive migration of radioactivity (Figure 4–6). Until now, the procedure has worked fairly well, and radioactivity has not reached the water table. This, however, is likely the result of a very deep water table and locally low precipitation levels rather than foresight in choosing Hanford as a disposal site.

As early as 1966, the National Academy of Sciences warned that disposal into soil could not be relied upon:

> [C]ontinuous disposals beyond this rate [an equilibrium where the rate of disposal does not exceed the rate of decay] could lead eventually to hazardous excesses of concentration, at which point the earth materials would no longer be a suitable disposal medium. (Lash, Bryson, and Cotton, 1974: 33)

Yet millions of gallons of low level liquids—much of it contaminated with plutonium—were dumped this way at Hanford. In 1972, the

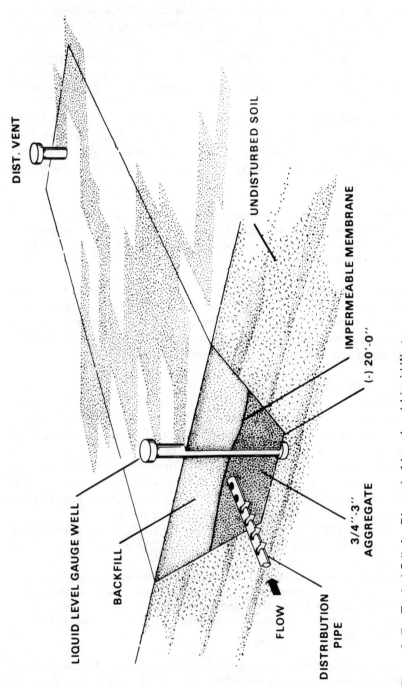

Figure 4–5. Typical Crib for Disposal of Low Level Liquid Wastes.

Source: AEC (1974b: II.1–89).

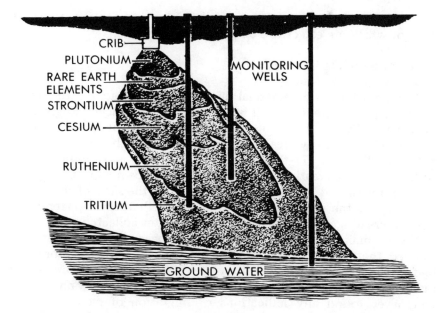

Figure 4-6. The Movement of Radionuclides Beneath a Typical Disposal Crib.

Source: AEC (1974a: fig. 12).

Atomic Energy Commission itself echoed the earlier findings of the National Academy, stating: "Soil columns and holding ponds and basins retain radionuclides. Radionuclides with longer half-lives can build up faster than the decay, which might result in unforeseen hazardous radioactive contamination . . ." (Lash, Bryson, and Cotton, 1974: 33). Indeed, such a problem did develop. In 1973, the AEC concluded that the concentration of plutonium at the bottom of one of the burial trenches might be great enough to cause a spontaneous chain reaction and even a low order nuclear explosion! (AEC, 1972: 7). A great deal of money was then spent to excavate the trench and remove and sequester the contaminated soil.

The practice of depending upon soil geochemistry to minimize the hazards of radionuclide dispersal into the ground is not always a desirable one. Arguing against the common generalization that soil is always an effective barrier against movement of radioactivity, the U.S. Geological Survey more recently stated:

> In some of the qualitative literature on waste disposal, a panacealike aura surrounds the term "ion-exchange." In such literature it is implied that when all else fails, ion-exchange processes will prevent movement of contaminants to points of water use. (Papadopulos and Winograd, 1974: 20)

Ion exchange is only one of a number of soil retention processes, none of which are effective 100 percent of the time.

Migration of Radioactivity at
Maxey Flats, Kentucky

The Maxey Flats commercial disposal site is located in the humid northern part of Kentucky. It is operated by the Nuclear Engineering Company (NECO), a subsidiary of the Teledyne Corporation. There, as elsewhere, solid wastes were dumped in trenches with unsealed bottoms (Figure 4−7) that were then covered and mounded over with earth, thus creating a "cap" with a high center. In theory, the high center should cause water to run off to the sides of the trench. In reality, as the material in the trenches settled, the caps tended to cave in, creating depressions in which rainwater collected.

NECO maintained that the rainwater was not a problem and claimed that plutonium in the waste would not migrate more than half an inch during one half-life of 24,000 years. NECO monitored radioactivity levels around and off the site on a semiannual basis and never found any indications of migration of radioactivity. A 1974 study by the state of Kentucky, however, yielded different results: plutonium contamination was found in surface soil, 90 centimeter deep soil cores, monitoring wells, and drainage streams (EPA, 1975: x). Evidently, plutonium had migrated deep into the ground and found its way off site at immensely faster rates than expected. (Several French geologists more recently described this as a "previously unheard of mobility of plutonium." DeMarsily et al., 1977: 525.) NECO ridiculed both the Kentucky report and one subsequently issued by the Environmental Protection Agency that reached the same conclusions, maintaining that the contamination detected was the result of plutonium transport by surface runoff water rather than by subsurface migration. According to the EPA, however:

> [I]t is possible to come to the reasoned conclusion that the [external] sources of contamination do not significantly affect the Pu data and do not rule out or lessen the importance of interflow and subsurface migration as potential pathways for Pu migration. (EPA, 1975: 27)

In other words, the EPA concluded that subsurface migration of plutonium had indeed occurred. Apparently, rainwater seeped through the trenches and into the ground beneath, carrying along the plutonium.

Recent research into plutonium transport mechanisms indicate that the radionuclide may have chemically combined with organic

Figure 4-7.   Disposal of Low Level Radioactive Wastes in a Shallow Burial Trench at the Maxey Flats, Kentucky, Disposal Site.

Source: Ron Garrison, Lexington *Herald-Leader.*

solvents used for decontamination purposes (Means, Crerar, and Duguid, 1978: 1477). A chemical compound of this type would be electrically neutral and would thus be unaffected by ionic retention processes in the soil.

After this incident, NECO began pumping water out of the trenches. The contaminated liquid was evaporated, and the residue reburied. The trench caps were also repaired. However, the trench bottoms are not and cannot be sealed. No one wants to consider the possibility of unearthing the wastes. The Maxey Flats site was closed in December 1977, but the problem of plutonium migration is by no means solved. The trenches will continue to be radioactive for many millenia and will evidently have to be pumped for the indefinite future. There is no organization that can give the appropriate guarantees that this will be carried out, and no arrangements have been made for paying the long-term costs of managing the site.

At the Store in Beatty, Nevada

Sometimes the human element causes embarrassing problems, as in the case of the Beatty, Nevada, low level waste disposal site, also operated by NECO. The Beatty site was known locally as the "store." If the townspeople had need for equipment or building materials, the site probably had them for sale, illegally, by site employees. This business continued for several years, until it was finally discovered and stopped. Tools, generators, plywood, and lab equipment, many radioactive, found their way into the town of Beatty. Most of the contaminated objects were eventually recovered, the implicated employees and manager fired, and the site license temporarily suspended (House of Representatives, 1976: 298).

In general, the record at other low level burial sites has been better. Even so, adequate management cannot make up for poor site location, and none of the presently active sites were chosen with optimum hydrogeologic ground characteristics in mind. Off-site contamination has also been detected at West Valley, Oak Ridge, and Idaho Falls, indicating that many, if not all, of the low level disposal sites are unsuitable. Currently, the Barnwell, South Carolina, site is the only one east of the Mississippi accepting low level waste—and on a limited basis—even though this is where the greatest volume of waste is produced.

Of the six commercial burial grounds, two have been shut down because of off-site radioactive contamination and poor operating practices, while a third has been closed because of an inability to acquire a new license that would allow site expansion. (The three sites are West Valley, Maxey Flats, and Sheffield, Illinois.) In July

1979, the governor of Nevada temporarily closed the low level waste disposal site at Beatty after a truck carrying radioactive waste to the site sprang a leak. (This was the second such incident in two months; on May 10, ten people were contaminated by radioactivity after a truck caught fire at the site.) The Beatty site was reopened several weeks later. In October 1979, the Richland, Washington site was closed temporarily for similar reasons. Thus, at the present time, none of the low level disposal sites are accepting waste on an unconditional basis. It seems likely that a new low level disposal site will have to be opened within the next several years, probably somewhere in the northeastern United States.[2] The discouraging experiences of commercial ownership led the Deutch Report to recommend that all low level disposal sites be owned by the federal government and operated through private contract. The poor operating records of some of the low level disposal sites also stresses the need for greater investigation of site suitability—as opposed to economic considerations, as has often been the case—of any facility proposed in the future.

THE MILL TAILINGS PROBLEM

There are over 140 million tons of uranium mill tailings at thirty-five sites in nine Western states, mostly in unstabilized or partially stabilized piles. The toxic radioactive gas radon diffuses from the radioactive materials in the tailings into the atmosphere, where it is borne away by the wind. Other radioactive materials, such as radium, are leached out by rain. As a result, radioactive contamination usually extends beyond the boundaries of the pile. In several cases, these piles can be found in close proximity to inhabited areas. One large unstabilized tailings pile is located only thirty blocks from the center of Salt Lake City, Utah. (Figure 4–8 is a photograph of an abandoned mill tailings pile in New Mexico.)

High levels of radioactivity have been found in water supplies located downstream from old mill sites, and until 1959, some uranium mills dumped tailings directly into nearby streams. This practice, which contaminated large portions of the Colorado River basin, was finally halted by the Atomic Energy Commission only under heavy pressure from the U.S. Public Health Service. Airborne radioactivity may also be quite high in the vicinity of tailings piles, and significant levels of gamma radiation have been measured at their surface. A

2. Indeed, according to a report in *Nucleonics Week* (1979j), the Rickano Corporation of Lyons, Kansas, a subsidiary of Southwest Nuclear, has submitted an application to the Nuclear Regulatory Commission to be allowed to use the old Carey salt mine under Lyons as a repository for the retrievable storage of low level radioactive waste.

Figure 4–8. Abandoned Mill Tailings Pile in New Mexico's Uranium Belt, left from an old milling operation owned by Phillips Petroleum.

Source: Rudi Schoenmackers, State of New Mexico Energy and Minerals Department.

study of radiological conditions at twenty-two locations where uranium mills have closed down found that:

> Based on the correlation observed between exposure to radon and other radium daughter products and incidence of lung cancer in uranium miners, the risk of incurring lung cancer is about double the normal to populations living in close proximity to the tailings. (DOE, 1978a: 25)

Until 1966, mill tailings were widely used as fill for building foundations.[3] Before the practice was halted, at least 5,000 buildings in Colorado, and several thousand in other states, were built on mill tailings. In Grand Junction, Colorado, alone, several thousand homes incorporated tailings in their foundations (Hollocher and MacKenzie, 1975: 49). The Colorado Department of Public Health estimated that 10 percent of the occupants in those homes received the equivalent of more than 553 chest x-rays per year due to gamma emissions from the tailings (Metzger, 1971). Until today, only about $6.5 million has been spent on the removal of tailings at fewer than half of the 700 sites requiring such remedial work (Carter, 1978b: 194). The actual cost may never be determined because complete records of buildings constructed with tailing fill were not kept.

Under the provisions of the 1954 Atomic Energy Act, the AEC decided that it had no jurisdiction over mill tailings. Today, a dispute exists over assignment of responsibility for the tailings. The federal government would prefer that the mill owners pay for disposal of tailings, but in many cases, the mills have been shut down, and the owners—at least those that can be found—are neither financially able nor willing to pay these costs. Congress has passed legislation intended to cover most of the cleanup costs, estimated to be in the range of $80 to $125 million (DOE, 1978a: 93), which is thought to be "optimistic" by some (Carter, 1978b: 193).

Not so obvious in this recitation of some of the more glaring failures of the domestic waste management program is the following point: None of these failures was caused by technical problems alone. To be sure, waste management technologies have failed, but, in virtually every instance, failure was the direct result of inappropriate economic or political decisions. For example, soft steel tanks were used for storage of liquid waste at the Hanford Reservation because of the cost and nonavailability of stainless steel. The Lyons, Kansas repository was ill-conceived and doomed from its inception

---

3. Free public access to the tailings was not restricted until 1972 (Hollocher and MacKenzie, 1975: 48).

because it was hastily prepared by the Atomic Energy Commission in an attempt to defuse a politically explosive situation. The mill tailings problem came about as a direct result of the AEC's refusal to regulate disposal of the tailing and not because of a technical inability to manage them safely. Lack of interest in and failure to anticipate the magnitude of the waste problem also contributed to the dismal record. These are not problems only of the past, however. As will be seen in the following chapter, which describes the current federal waste management program, there are many lessons from the past that are still relevant today.

※ *Chapter 5*

The Present Radioactive Waste Management Program

Until the Atomic Energy Commission was abolished, there was not, properly speaking, a coherent, comprehensive program for managing radioactive wastes. If the AEC had a policy with regard to waste disposal, it was one of deferring a solution to the indefinite future. As a result, waste management was effected through a series of short-term "technical fixes," many of which proved inadequate virtually upon implementation. In addition, these inappropriate fixes were always put into practice with little or no thought given to critical nontechnical, or "institutional," problems that might develop, such as human error, political resistance, or bureaucratic bungling. This omission inevitably exacerbated the impact of specific failures, for not only were the AEC's waste managers made out to be hopeless technical incompetents, but they were also seen as politically arrogant and insensitive. In this way, the AEC created a climate of distrust and contempt that persists to this day.

Unfortunately, the current waste management program of the Department of Energy is subject to some of the same weaknesses that plagued waste management under the AEC, for the department strongly believes that proper waste management is primarily a technical problem with a very small nontechnical component. Programs have been initiated with little or no public consultation. Ill-conceived projects, designed, in our view, to ease pressure on a beleaguered nuclear industry, are being pushed forward. Extensive plans are being laid for interim storage of spent reactor fuel. As a result, a great deal of political and societal resistance to waste management projects has developed. This is not to imply that the federal government is completely insensitive to institutional problems, however. The report of the Interagency Review Group on Nuclear Waste Management (IRG, 1979), a committee of representatives from various federal agencies appointed by President Carter in 1978 to draw up a comprehensive program for waste management, gives considerable thought to the institutional aspects of a waste management program. Whether the conclusions of this report will be integrated into DOE's program remains to be seen.

The current domestic radioactive waste management program described in this chapter is still in a state of great flux. What is described here may not necessarily become final policy. For example, a decision to reprocess spent fuel could lead to a policy of long-term fuel storage as opposed to disposal. A change in government policy on this issue could cause a complete reversal in the direction of the program. This chapter also discusses some of the technical and institutional problems that stand as obstacles to successful implementation of the government's program. The status of foreign waste management programs is discussed in Appendix C. It is important to note that, although several European governments are somewhat more advanced than the United States with regard to waste solidification technology, none is any further advanced in the area of permanent waste disposal.

THE CURRENT FEDERAL WASTE DISPOSAL PROGRAM

In 1975, after cancelling the Retrievable Surface Storage Facility, the Energy Research and Development Administration decided to resume active investigation into the geologic isolation concept. This is still the direction of the program today. The current waste management program may be conveniently divided into three parts—interim spent fuel storage, the National Waste Terminal Storage program, and the Waste Isolation Pilot Plant project.

Interim Spent Fuel Storage

According to the schedule established by ERDA in 1975, a commercial high level waste geological repository was to be operational by 1985. Reprocessing of spent fuel was expected to be an established practice, and thus, spent fuel would remain at reactor sites only for a limited period of time. However, the schedule for a repository has slipped to 1990 or later, and with the indefinite deferral of reprocessing, spent fuel is piling up at reactor sites. By 1983, some reactors may have to cease operation for lack of space to store temporarily the entire core inventory of fuel in the event of repairs or an emergency situation. One logical solution to this impending problem would be the construction, by the reactor operators themselves, of more on-site spent fuel storage capacity. The electric utilities, however, are quite opposed to this idea, desiring neither the expense nor the indefinite responsibility for storing the waste.

The Department of Energy, on the other hand, would like to see the reactor operators rerack and compact spent fuel pools and shuffle spent fuel between sites wherever possible. (However, DOE is not favorably disposed toward construction of additional storage capacity.) Depending upon how successful this strategy is, it is estimated that 500 to 3,000 metric tons of spent fuel will require interim off-site storage by 1983 (*Nucleonics Week*, 1979i: 3). To handle this excess, the DOE proposed construction of a federally owned, privately operated away from reactor (AFR) spent fuel storage facility by 1983. Persistent failure to achieve congressional funding for the AFR concept has pushed the date for an operational facility to 1984–1985. As a stopgap measure, DOE is expected to decide to store spent fuel in one of the spent fuel pools at the three idle reprocessing facilities at Morris, Illinois, West Valley, New York, and Barnwell, South Carolina. However, this idea may also run into trouble. New York State officials and congressmen for the most part oppose future use of West Valley for nuclear activities (although the preliminary agreement reached between the state and DOE included a provision for utilization of the spent fuel pool—see Chapter 4). Both General Electric and Allied-General, respectively owners of the Morris and Barnwell facilities, have stated that they have no interest in having their facilities used as AFRs.

The National Waste Terminal
Storage Program

The National Waste Terminal Storage (NWTS) program was established to:

> [P]rovide terminal storage facilities for commercial radioactive waste in geologic formations at multiple locations in the United States [and] to provide the capability to safely dispose of any commercial radioactive waste that must, according to federal regulations, be delivered to a federal repository for terminal storage. (OWI, 1978: 3)

In 1975, ERDA contracted with the Union Carbide Corporation to manage the NWTS program. Union Carbide organized the Office of

Waste Isolation (OWI) at Oak Ridge, Tennessee, to run the program.[1] OWI was directed to locate six sites suitable for high level waste repositories. However, even surveying for the six sites has not been an easy task. When OWI initiated its search, of the research done on potential disposal media, that into salt was the most advanced. As a result, site suitability surveys were, for the most part, limited to states with salt formations within their borders. Among these were New York, Michigan, Texas, Louisiana, Mississippi, Ohio, Utah, and Colorado. Most of these states have since proven to be strongly opposed to the notion of high level waste disposal within their borders. For instance, a survey of the Salina salt formation in Michigan was initiated without proper notification of state and local authorities, and in a near repeat of the Lyons debacle, the governor of that state asked ERDA to cease its investigations. Until March 1979, Governor Carey of New York was still adamantly opposed to the establishment of any new waste management facility in his state and refused to allow DOE site surveyors access to the New York portion of the Salina Basin. In March 1979, New York and the Department of Energy reached an agreement over future use of the West Valley site (see Chapter 4), which included the possibility of high level waste management activities at the site. Whether the agreement includes permission for DOE to conduct site suitability surveys is not clear. In addition, events at Three Mile Island placed the agreement in jeopardy, and its status is uncertain at this writing. An in situ electric heater experiment in a northern Louisiana salt dome has incurred the wrath of state representatives, apparently raising fears that a waste repository might be constructed there; and pending site surveys in eastern Texas, the Texas panhandle, and Ohio are evidently causing some concern among the populace. (Bedded salt surveys are still underway in Utah and Colorado.) In some instances, DOE has also been unable to arrange meetings with appropriate state officials to discuss the possibilities of site suitability investigations (Heath, 1978).

These problems have caused considerable slippage in the NWTS program. As recently as 1978, DOE had hoped to identify a suitable repository site in bedded salt by 1980 but, more recently, has largely abandoned its efforts to locate an early site in this medium and is hoping instead to choose a dome salt site by March 1981. As a result of site-surveying problems, a bedded salt site cannot be chosen before 1983 (*Nucleonics Week*, 1979e: 2).

1. The Office of Waste Isolation was reorganized in 1978 as the Office of Nuclear Waste Isolation, under the management of the Battelle Memorial Institute in Columbus, Ohio.

Because of these "political" problems and the technical uncertainties associated with salt disposal (described in a previous chapter), investigations into other geologic media have been intensified. In particular, research on granite, begun at the Nevada Test Site in 1976, and on basalt, begun at the Hanford Reservation in 1977, has been speeded up.

The Nevada Test Site is already highly contaminated with radioactivity as a result of some 500 nuclear tests—the majority underground—conducted there. DOE is currently characterizing the southwest corner of the site as a potential commercial high level waste repository. In 1980, DOE plans to encapsulate nineteen spent fuel elements in stainless steel and to store them in various configurations at the Nevada Test Site in a program expected to last three to five years (*Nucleonics Week*, 1979b: 1). Of the nineteen canisters, eleven will be emplaced in a granite tunnel, originally constructed for a nuclear weapons test, 1,400 feet below the surface. One canister will be placed in an instrumented concrete silo (the so-called "Stonehenge concept" according to DOE), and four others will be placed in dry wells. The remaining three canisters will be used as replacements. Politically, the Nevada Test Site may be an ideal location for a repository: it is federally owned and dedicated to nuclear uses, the surrounding population density is quite low, and the citizens of Nevada are generally inclined to support nuclear projects. Whether they will accept a waste repository at the test site remains to be determined.

DOE hopes to select a basalt repository site by September 1981 (*Nucleonics Week*, 1979e: 2). It will probably be located in Oregon, Idaho, or Washington. At the present time, the Rockwell Hanford Company is pursuing a program of tests at Hanford in order to acquire badly needed information about the characteristics of basalt. In July 1978, Rockwell began construction of a test facility in a basalt outcrop at Hanford. In late 1979, the company intends to place waste-simulating electric heaters in the facility, located about 100 feet underground. This will be followed in 1980 by emplacement of twenty-two spent fuel assemblies in the facility (*Nucleonics Week*, 1978a: 1). Rockwell reportedly hopes to have an operating repository at Hanford by 1989. The facility will be located at a depth of 3,000 feet, will cover sixteen square miles underground and two to four square miles on the surface, and will cost over $1 billion (*Nucleonics Week*, 1978a: 1). It has also been proposed that high level waste be buried at Hanford in tunnels cut into hills (NAS, 1978: 108) or in the reservation's dry, arid soil (Winograd, 1976), but neither of these two proposals is under serious consideration at the present time. Hanford is an attractive site for a waste repository because, like the Nevada Test Site, it is federal property and has gained general acceptance by Washington citizens as a nuclear facility. There is, however, growing opposition to siting a repository at Hanford. Furthermore, the reservation's proximity to the Columbia River must raise some serious questions as to Hanford's suitability as a site for a waste disposal facility.

As with other nuclear-related projects, the NWTS schedule has slipped badly and is likely to continue slipping. The current projected date for operation of the nation's first commercial high level waste repository is 1988−1989. However, at least ten years will be required for licensing and construction once a site is chosen, and so, with site selection postponed until 1981 at the earliest, a repository cannot be operating much before 1991. Because more problems will inevitably arise in locating a suitable site, and with licensing procedures bound to take longer than expected, it does not seem unreasonable to predict that the NWTS repository will not be in operation before the period 1995−2000.

The Waste Isolation Pilot Plant

The WIPP Program. The Waste Isolation Pilot Plant (WIPP) program was established as a result of the cancellation in 1972 of the Lyons, Kansas, project. The Lyons project itself was initiated in 1969 in response to the AEC's agreement with Senator Frank Church to remove all transuranium-contaminated waste from Idaho by 1980. WIPP, too, was intended to fulfill this promise. Although in political terms the Lyons project was a clear failure, it neither confirmed nor disproved the feasibility of salt disposal. Accordingly, in 1973 a search was begun by the AEC, U.S. Geological Survey, and Oak Ridge National Laboratory to locate a new site for a salt repository. In 1974, investigators settled on a site to the east of Carlsbad, New Mexico, located in the Salado salt beds of the Delaware Basin, which, in turn, is part of the larger Permian Basin whose salt beds extend into Texas, Oklahoma, Kansas, and Colorado. The site was thought to be a good one for several reasons: it was isolated; it supposedly was similar to the Kansas salt beds; and it was located in New Mexico, a state that has consistently supported nuclear-related projects. However, intensive site surveying for WIPP did not begin until 1975, after the demise of the Retrievable Surface Storage Facility. Soon thereafter, the Energy and Research Development Administration contracted with Sandia Laboratories of Albuquerque, New Mexico, to manage the WIPP program.

As presently planned, WIPP is to be a mined repository in the 2,000 foot thick Salado formation (Figure 5—1). It will have disposal levels at 2,100 and 2,700 feet below the surface. The exact types of waste to be buried in WIPP are not yet clear, however. As originally conceived, WIPP was to contain only transuranium-contaminated military waste. In 1977, it was suggested that high level defense waste might also be interred in the repository. More recently, the Deutch Report (DOE, 1978a: 17) proposed that up to 1,000 spent fuel assemblies—about 500 metric tons—be emplaced in a section of WIPP in a retrievable manner so as to "demonstrate" the scientific and technical feasibility of geologic isolation in bedded salt. The spent fuel demonstration has become an official part of the WIPP program, but, for political reasons described below, has engendered serious congressional opposition. In addition, the WIPP site appears to be seriously flawed from a technical point of view; some of these problems are described in the following section. If constructed as described in the draft environmental impact statement prepared by Sandia Laboratories for the Department of Energy, WIPP will contain a one hundred acre area for disposal of all transuranium-contaminated waste currently stored in Idaho, a twenty acre area for salt research and development, and a second twenty acre for the permanent disposal of as many as 1,000 spent fuel assemblies (DOE, 1978b: 1—2).

Because WIPP is a defense-related facility, legislative funding for the project originates with the U.S. House of Representatives Armed Services Committee. That committee has refused to authorize funding for WIPP if the project is to be licensed by the Nuclear Regulatory Commission and used for disposal of commercial spent fuel. Some members of Congress apparently fear that the environmental impact statement required for the nondefense portions of WIPP could set a precedent for other defense-related facilities by allowing citizen challenges to defense projects. None of the other House committees that could fund WIPP has shown any interest in doing so. As a result, the Department of Energy was unable to procure funding for WIPP for fiscal years 1979 or 1980. Although DOE still strongly favors including the spent fuel disposal project in WIPP and having the facility licensed by the NRC, the department may be wavering in its commitment to WIPP. The report of the Interagency Review Group on Nuclear Waste Management suggested that a spent fuel project such as that proposed for WIPP might be unnecessary; a similar project could be implemented as an early stage of a full-scale repository (IRG, 1979: 57). One DOE official has stated that the

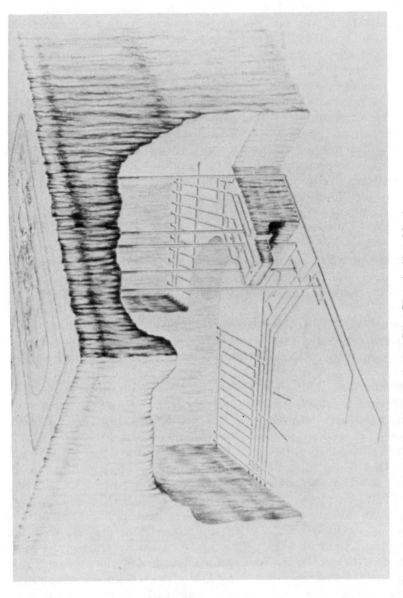

Figure 5–1.   Artist's Conception of the Waste Isolation Pilot Plant in New Mexico.

Source: DOE.

delay or outright cancellation of WIPP would not seriously affect the schedule for a high level waste repository (*Nucleonics Week*, 1979h: 6). The final decision to continue or cancel WIPP has been left to President Carter.[2]

If the WIPP project proceeds as planned, its operational date is uncertain. Originally scheduled to begin accepting waste in 1984, the opening data has now slipped to a still unrealistic 1985–1986. Because of the inevitable construction and licensing delays, however, WIPP is not likely to be operating much before the late 1980s. In that case, the project may prove unnecessary, because the transuranium-contaminated wastes destined for disposal in WIPP could easily be buried in the first operational high level waste repository. The estimated cost of the WIPP project has increased from $100 million to $500 million (*Nucleonics Week*, 1979c: 1), and further delays could easily increase this to $1 billion. Abandoning WIPP now and dedicating WIPP funds to high level waste repository research would make a great deal of sense. In light of the political problems described above and the technical uncertainties and technical weaknesses of the WIPP site, described in the following section, the project's future is not a bright one.

Technical Uncertainties about the Waste Isolation Pilot Plant.   Despite widely voiced reservations about the suitability of salt as a disposal medium, the Department of Energy is proceeding with the Waste Isolation Pilot Plant. Moreover, available evidence suggests that the WIPP site is a technically unsuitable location for a waste repository, even if there were no generic difficulties with salt, for reasons discussed below:

1. *Salt dissolution features are present in and adjacent to the WIPP site.*   Large diameter dissolution features called "breccia pipes" are frequently found associated with bedded salt. A breccia pipe is a cylindrical region that extends from the surface, where it may appear as a rubble-filled depression, into and through deep sedimentary layers, including salt beds. Figure 5–2 shows a diagrammatic illustration of a breccia pipe. It may be 100 to 500 meters in diameter, filled with loosely packed material through which water is able to flow. Although the mechanism by which a breccia pipe forms is unknown, one credible theory has been proposed: water from an aquifer under

2. According to a report in *Nucleonics Week* (1979k), DOE has abandoned plans to license WIPP and to conduct research and development on the disposal of spent fuel at the site. The ultimate purpose and/or fate of WIPP, however, is still in the hands of President Carter.

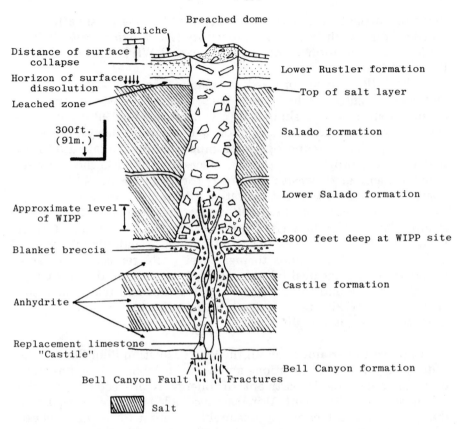

Figure 5-2. Diagrammatic Illustration of a Composite Breccia Pipe originating from dissolution in salt above the Bell Canyon aquifer under the WIPP site. Not all breccia pipes will show this combination of all features.

Source: Anderson (1978).

the salt bed can leak through the impermeable layer separating the aquifer from the salt, dissolve salt, and become dense (relative to fresh water). It then sinks back into the aquifer, and fresh water replaces the denser salty water. The fresh water picks up salt, sinks, and is replaced by more fresh water. Thus, a density-driven convective flow is established, capable of eating a vertical chimney through the salt bed. Nonsalt layers can also be penetrated through the action of complex types of bacteria whose fossilized remains have been found in breccia pipes in the Delaware Basin (Montague, 1979: 3; Kerr, 1979b: 603). The breccia pipe problem is of special concern for WIPP because a half-mile diameter sink has been discovered about 1,650 feet below the site (Figure 5–3). The nature of the sink has not been determined, but if this sink is a breccia pipe, it could provide a pathway for surface water to enter the WIPP repository at some future date, resulting in radionuclide migration.

Another potential concern is the presence of circulating groundwater to the west of the WIPP site:

> A front of groundwater that is dissolving the salt formation at a rate of four to six miles per million years is moving towards the proposed site from the west. . . . The dissolution front is 3 to 10 kilometers [2 to 6 miles] from the boundaries of the study area. (CERCDC 1978a: 188)

Subsurface salt dissolution has also occurred to the northeast of the WIPP site. Ten thousand years ago, the Southwest was much more humid than it is now. Future changes in climate of this type and potential human intervention could alter the groundwater regime so as to increase the dissolution rate. With the dissolution front so uncomfortably close to the WIPP site, it may be impossible to guarantee the rule of thumb isolation period of 250,000 years.

In fact, groundwater flow patterns in the WIPP region have not been well studied. Should the repository fail because of water intrusion, it would be desirable that groundwater flow paths be long enough to assure decay of the hazardous radionuclides before they reach the environment. The necessary investigations into this question have yet to be initiated.

2. An anticline has been found under the WIPP site.   An anticline is a fold in a rock stratum. It is usually caused by tectonic forces acting upon the rock. In salt beds, however, the presence of such a fold may indicate deep dissolution of salt by circulating groundwater (Anderson, 1978: v).

The first WIPP site was abandoned in 1975 when investigators drilled into a pocket of geopressurized brine. It was later discovered that the brine was associated with an anticline, and this was deemed sufficient reason to abandon the site and relocate operations to a new site several miles southwest of the old one.

It appears that there may be an anticline under the present WIPP site (Figure 5–4). If the presence of an anticline was unacceptable at the original WIPP site, it is not clear why it is acceptable at the present one.

3. Faults penetrate the WIPP site.   Studies by Sandia Laboratories have found three intersecting fault lines that run though the part

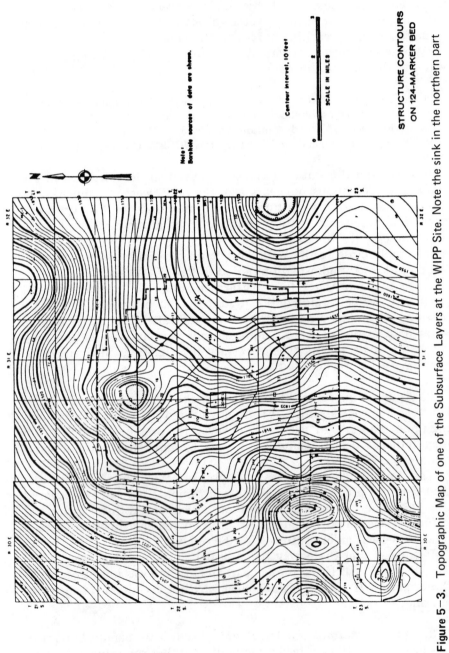

Figure 5–3. Topographic Map of one of the Subsurface Layers at the WIPP Site. Note the sink in the northern part of the site. It is located about 1,650 feet below the surface.

Source: Sandia (1978: fig. 4 4–7).

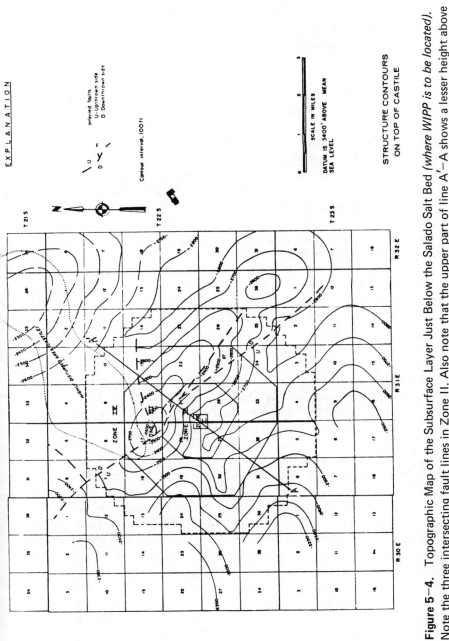

Figure 5-4.   Topographic Map of the Subsurface Layer Just Below the Salado Salt Bed *(where WIPP is to be located)*. Note the three intersecting fault lines in Zone II. Also note that the upper part of line A'-A shows a lesser height above sea level than the lower part of the same line. This is the anticline referred to in the text.

Source: Sandia (1978: fig. 4.4-6).

of the WIPP site intended for waste disposal (Zones I and II in Figure 5–4). As the topographic map shows, the vertical displacement along these faults indicates that the site may be tectonically active. As previously noted, inactive faults could allow groundwater to enter a waste repository if, at some time in the future, they were to become active.

4. *Low level seismic activity near WIPP could threaten repository integrity.* The Central Basin Platform, the area in which the WIPP site is located, is generally considered to be free of earthquakes. Nonetheless, since August 1966, there has been a series of minor earth tremors originating about fifty miles southeast of the site. According to the U.S. Geological Survey, an assessment of the regional seismicity of the Central Basin Platform is required because of "a series of low-level events [earthquakes] which may be related to water flooding of oil wells in the vicinity. The seismicity is significant for any hazard it may pose to the nearby Waste Isolation Pilot Plant . . ." (CERCDC, 1978a: 150).

Water injection is a common method of extending well production lifetime and, if ever allowed in the immediate vicinity of WIPP, could affect the repository. Thus, control will have to be exerted for very long periods of time over activities in the area surrounding WIPP. At present, control extends over a buffer zone of only some thirty square miles.

5. *Mineral exploitation around the WIPP site could threaten repository integrity.* The Nuclear Regulatory Commission has recommended that waste repository sites "not offer an attractive target for future generations seeking natural resources" (NRC 1977c: 5). Yet the WIPP site offers just such an attractive target for the current generation of resource seekers. It lies over a petroleum reservoir estimated to contain between 40 and 490 billion cubic feet of natural gas, 38 million barrels of crude oil, and 5.7 million barrels of distillates.[3] The U.S. Geological Survey estimates that the WIPP site contains some 350 million tons of high grade lanbinite potash worth over $1 billion (*Nucleonics Week*, 1979d: 2, Bureau of Mines, 1977). Lanbinite potash makes exceptionally good fertilizer. At the present time, southeastern New Mexico is the largest domestic source of potash and is likely to become an increasingly important source as the worldwide demand for fertilizer increases. Furthermore, salt itself may be an important resource. Hydrocarbon and potash leases within the WIPP site are shown in Figures 5–5 and 5–6.

3. Note, however, that at present domestic consumption rates, this resource would only yield a few days' supply of hydrocarbons.

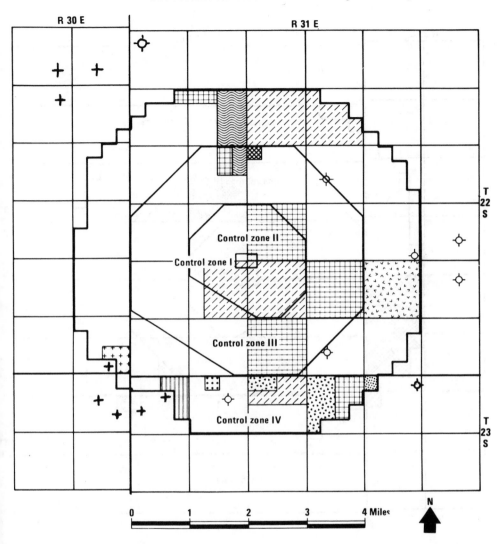

-◇- Abandoned wells

+ Deep producing gas

Figure 5–5. Oil and Gas Leases Within the WIPP Repository Site. Also indicated are abandoned and deep gas producing wells. *(Leases are shaded areas.)*

Source: DOE (1979b: 8–11).

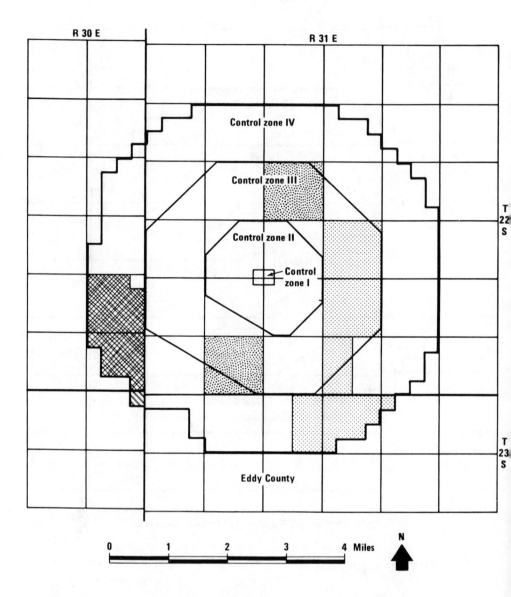

Figure 5-6.   Potash Leases Within the WIPP Repository Site. *(Leases are shaded areas.)*

Source: DOE (1979b: 8-10).

These resources are likely to be tapped. Gulf Oil, which holds several federal leases on the proposed WIPP site, has indicated that it will sue if not allowed to extract the petroleum (*Nucleonics Week*, 1979d: 2), and ultimately, the potash is likely to be extracted, too. Oddly enough, WIPP has already been affected by the resource problem. The site is located in the southwestern corner of the Permian Basin, an area of heavy petroleum exploration and extraction (Stevens, 1978: A18). In its initial search for a suitable site, the Energy Research and Development Administration tried to find a spot meeting the condition that it be free of petroleum boreholes within two miles of the site center. The search was unsuccessful. Rather than give up, ERDA simply changed the required borehold-free radius to one mile. (Figure 5–5 shows all drill holes within and adjacent to the WIPP site.) Eventually, the WIPP site is likely to be disrupted by people seeking natural resources. This, and the other problems outlined in this section, strongly suggest that the WIPP site should be abandoned as unsuitable for a waste repository.

THE HUMAN PROBLEMS

Too often, those who plan programs of great technological complexity neglect to consider the role that human behavior can play in frustrating program objectives. Yet it is frequently aberrant or careless human actions that cause weaknesses and failure in complex and important technical programs. This becomes an issue of concern in two areas of radioactive waste management: (1) prevention of human intrusion into stored or buried waste at some time in the near or distant future, and (2) prevention of what is here called "institutional failure"—that is, understanding and control of collective human error in the planning and implementation of a disposal technology.

The Problem of Human Intrusion

One of the more obvious causes of the long string of failures in the U.S. radioactive waste management program has been human error—failure to check liquid waste tank levels, for example, or failure to assess adequately the technical suitability of a proposed repository site. In the absence of a strong and continuing commitment to safe radioactive waste management and disposal, such failures are likely to recur. In the future, closer regulation, technical redundancy, and frequent inspection of facilities may be able to minimize the frequency, magnitude, and consequences of such failures. However, deliberate attempts to disrupt waste management facilities, such as might be organized by terrorist groups, probably cannot be prevented. Such attempts would have a not insignificant probability of success, with serious consequences to the public. Nonetheless, over the short term, accidental and deliberate failures can best be controlled by institutional means. Over the longer term, once effective institutional control has been lost, human intrusion into radioactive waste storage and disposal facilities must be considered a potential cause of radioactivity release.

If radioactive wastes are stored in surface facilities, as has been proposed in the past, and institutional control is lost, the chances of future disruption by human societies is quite large. On this subject, Dr. Gene Rochlin of the University of California at Berkeley has written: "Construction of a large concrete mausoleum . . . would almost guarantee that concerted efforts would be made to breach it by intelligent, but uninformed life" (1977: 27). To give one example of this, the Pyramids have been subject to human intrusion repeatedly over thousands of years, despite widespread superstition and fear about the consequences of such break-ins. Certainly, a surface waste storage facility would be similarly vulnerable.

However, neither does geologic isolation preclude human intrusion, accidental or deliberate. The Department of Energy has indicated that, after decommissioning, a repository site will be clearly marked with "permanent" warning signs. While a warning of this type might be satisfactory for hundreds or even thousands of years, it might just as easily cease to be effective within decades. Even if the markers were evident to investigators of the future, their meaning might not necessarily be clear. Rochlin writes: "We cannot assume that . . . a society . . . will be able to decipher a message it cannot read. Indeed, the presence of such an indecipherable message would only arouse additional interest" (1977: 27). To support this assertion, he points out that between the discovery of the Mycenaean tablets at Knossos and their deciphering, there elapsed a period of fifty years. Rochlin adds: "The 3500 years that have elapsed since the inscriptions were made is only about one-seventh of the half-life of plutonium–239. Yet, almost all the information about the culture and language of Minos had been lost" (1977: fn.60).

A future society's search for natural resources might also cause breaching of respository integrity. Salt, for example, is commonly associated with oil and gas and is itself a valuable mineral. Drilling for resources, although not likely to cause widespread radioactive contamination, could result in substantial radiation exposure to the drilling crew (DOE, 1979b: 9–125). A borehole into a repository could open the way to substantial intrusion of surface waters followed by significant waste transport. Rochlin adds a further caveat

regarding salt: "Interesting geologic formations such as salt domes...
are likely to draw attention" (1977: 27). It is sobering to realize that
of the approximately 150 shallow salt domes known to exist in the
Gulf Coast region, fully two-thirds have been breached by drilling
(Johnson and Gonzales, 1978: 174). While a basalt or granite repository may not be in an "interesting" formation or an area associated
with mineral resources, it is likely to be located in or close to an
aquifer. This water source could be exploited for irrigation purposes
or, conversely, could be affected by irrigation in the surrounding
hydrologic basin. In either instance, groundwater flow patterns could
be so seriously altered as to speed up migration of radioactive waste
into the environment.

The consequences of human intrusion could vary greatly, depending on the nature of the breach. As noted above, drilling into a waste
repository would probably cause only very localized contamination.
Alteration of regional hydrology followed by the introduction of
radioactive waste into water sources could cause significant radiation
exposure and health hazards under some circumstances. Still, the
effects of repository breaching would be unlikely to reach beyond
distances of tens or hundreds of miles. This suggests that siting of
repositories in arid, isolated locations, in addition to adherence to
other important site selection criteria, could minimize the consequences of human intrusion, at least over the period of greatest
hazard. Much beyond several hundred years, however, a "favorable"
geographic location is not likely to provide a barrier to human intrusion. For the longer time period, locating repositories in "uninter-
esting" rock devoid of mineral resources may be the best way to
ensure against disruption of the facility.

The Problem of Institutional Failure

Effective implementation of a safe radioactive waste disposal system depends upon the competence and persistence of the implement-
ing institutions. Failure in the planning and implementation process
could render ineffective even the most promising disposal technology, and the public response to such failure could severely limit the
freedom to implement similar systems at a later date.

Institutional failure could be the result of one or many poorly
made decisions—with respect to site selection, site geology and hydrology, repository construction, waste solidification, encapsulation
and emplacement, and repository backfilling and sealing—that might
lead to inferior design and excessive leakage of radionuclides from
the facility. Failure could also be caused by undue haste in imple-
menting a "solution." Finally, political pressures aimed at bolstering

the sagging nuclear industry could impel institutions to "demonstrate" a "solution" even though flawed and despite contrary technical evidence. A member of the California Energy Resources Conservation and Development Commission, Emilio E. Varanini, III, has identified these concerns:

> It is our fear that the promoters of technology will brush aside the cautious and thorough approach which the development and assessment of nuclear waste technology requires and instead set politically motivated time schedules, rush ahead with engineering approximations instead of sound technical foundation, and commit us to a hasty program of "demonstration." Such a program would have as its principal objective not finding a safe, technically sound solution, but the creation of a climate to allay public fears and to cosmeticize an issue which American and even world public opinion acknowledges as a major impediment to our nuclear future. (1978: 3)

The consequences of institutional failure are not difficult to imagine. Should the first attempt to implement a geological repository prove less than successful, there could result great political and public pressure to abandon the project. The freedom to implement a second facility would be severely restricted as a result. There would of necessity then be a reversion to dependence upon surface storage facilities whose long-term reliability and security are, as we have seen, open to serious question. Ultimately, as a result of increased costs and lessened interest, pressures would develop to further defer permanent waste disposal. The ensuing proliferation of surface facilities could present a significant health hazard to future generations.

It could be argued that institutional failures should not be viewed as a source of concern for the future, but the failures of the past militate against such an attitude. According to a consultant to the Nuclear Regulatory Commission, radioactive wastes were ignored for the greater part of the Atomic Age because, "Lacking the sex appeal of reactor development and the pork barrel quality of other segments of the fuel cycle, waste management became organizationally and operationally, a residual category" (Metlay, 1978: 2). As a result, the careful management required for safe handling of the radioactive wastes was not forthcoming. Thus, reprocessing waste tanks leaked at Hanford; plutonium migrated at Maxey Flats; the West Valley reprocessing plant, ultimately a technical failure, had to be abandoned; and the Lyons repository failed to survive informed scrutiny. These failures and others testify to an institutional reluctance to deal with a nonglamorous but very serious problem in a forthright and competent manner.

A similar problem has developed within the Department of Energy, the institution now charged with management of the nation's radioactive wastes. This is due, in no small part, to the department's dual role in promoting nuclear energy and disposing of radioactive waste. Because waste disposal has become an obstacle to further expansion of nuclear power, DOE has a stake in seeing the problem solved as soon as possible. It appears motivated, to a large degree, by the pressures described by Commissioner Varanini. Under these circumstances, serious failures in the program are very likely to occur.

What, then, is to be done? Institutional failure can never be precluded completely, but mechanisms exist that, if applied to the task of organizing the waste management program and the institutions responsible for its implementation, could guard against failures that might otherwise develop.

The National Environmental Policy Act (NEPA) is one such mechanism. NEPA requires that any major policy decision, such as the development and construction of a repository, be assessed in order to determine its environmental impact and be compared to all reasonable alternative policy decisions. A waste disposal program that adhered strongly to NEPA regulations, based upon the development of repositories within several geologic media, could provide the flexibility that would minimize the impact of failure of a single facility. Extensive involvement of the public in repository siting and development and other policy decisions could also go a long way toward guarding against the types of institutional failures that have happened so frequently in the past.

At the present time, there exists, in our view, a serious conflict of interest within the Department of Energy between the promotion of nuclear energy and the disposal of nuclear wastes. Elimination of this conflict of interest, for example through the establishment of a competent, independent waste management authority, could work toward preventing institutional failures in the future. Recommendations toward these ends are presented in Chapter 6.

Chapter 6

Requirements for a Successful Program — Conclusions and Recommendations

The radioactive waste problem that we face today has been many years in the making. Over thirty years ago, the federal government commenced "temporary" storage of highly radioactive liquid waste in steel tanks in the belief that disposal was best deferred until a later date. Save for what has been lost through leaks, the wastes remain in those, or similar, temporary tanks. Almost twenty years ago, electric utilities began placing spent fuel assemblies in "interim" storage pools, assuming that the fuel would remain in storage for a limited time. The spent fuel remains, for the most part, where it was originally placed. Ten years ago, the Atomic Energy Commission promised that all transuranic waste in the state of Idaho would be removed and placed in a repository by 1979. That waste has not been moved. Today, the Department of Energy promises final waste disposal by 1990 or 1995 and at the same time emphasizes the need for "interim" away from reactor spent fuel pools to relieve pressure on the reactor operators. In view of the "history of unbroken failure to produce an acceptable method of waste disposal," as one public official has put it (Speth, 1979), there is little reason to be optimistic about this prediction. Given the undeveloped technical state of the various waste management and disposal options and the developing political and social problems of the current federal program, is it likely that successful waste disposal will be achieved by the end of the century? One can have little confidence.

From a technical point of view, the means needed to accomplish this goal are likely to be fairly well developed within the next decade or two. However, dealing with the political and societal requirements needed to ensure success will be much more difficult. It is our judgement that without major changes in the current program—particularly with regard to the nontechnical requirements—an acceptable means of disposal will be extremely difficult to develop and to implement. The bases for this conclusion are presented in this chapter. We begin with a brief review of the salient points of the problem. We then discuss what we believe to be the technical, political, and societal requirements of a successful disposal effort and evaluate the current program in light of these requirements. Finally, we present in the form of recommendations those changes we believe must be made in the current program in order to ensure its ultimate success.

REVIEWING THE PROBLEM

The Radioactive Waste Problem is a Large One and Growing Larger

A great deal of radioactive waste is presently buried or stored at numerous locations around the United States. Some 17,000 spent fuel assemblies—the product of only about 450 reactor years[1] of plant operation—are stored temporarily in spent fuel pools. This number is currently increasing by about 4,000 each year. About ten million cubic feet of high level liquid and solid waste produced by the reprocessing of plutonium for defense purposes resides in aging steel storage tanks at Hanford, Savannah River, and Idaho Falls, and some 600,000 gallons of high level waste are stored at the now abandoned West Valley plant. There are sixty-five million cubic feet of low level wastes—of which fifteen million contain transuranic nuclides—in shallow burial or storage at various government sites, and another sixteen million cubic feet of commercially generated low level radioactive materials are buried in six licensed waste facilities (of which three are permanently closed). Approximately 140 million tons of radioactive uranium mill tailings have been left in unstabilized or partially stabilized piles. Some of these tailings, used in construction, contaminate thousands of public and private buildings. Finally, many hundreds of obsolete, radioactively contaminated buildings at government defense facilities await decommissioning, dismantling, and disposal.

Even though these quantities are already quite impressive, they will inevitably increase. This nation's nuclear power program is still slated to expand, and as reactors now planned or under construction

---

1. One reactor year is the equivalent of one reactor operating for one year.

are completed (Figure 6–1), they will begin to produce not only electricity but also ever-increasing quantities of radioactive waste. By 1995, the annual production of commercial spent fuel could equal today's total inventory. By the century's end, the United States may have as many as 300,000 spent fuel assemblies in temporary storage— some 100,000 metric tons (Figure 6–2)—or seventeen times the number in storage today. The uranium requirements of a growing nuclear program could result in the production of up to one billion tons of uranium mill tailings. Low level waste inventories will run into the hundreds of millions of cubic feet. Quantities of this magnitude will greatly strain the storage capacities of temporary facilities and underline the pressing need for development of a permanent means of disposal.

The Radiological Hazard of These Wastes is Significant

The radiological hazard posed by these wastes is of potentially great significance. For example, the plutonium–239 contained in the radioactive spent fuel discharged by one reactor after one year of operation would be sufficient to cause fatal lung cancers in the entire population of the United States if dispersed as fine particulates and inhaled.[2] While we do not suggest that such material would be released to induce such damage, even the escape of small amounts of radioactive waste into the environment may result in a perceptible increase in the number of cancer deaths in a population. Or radionuclides, free in the environment, may be concentrated in marine organisms and terrestrial animals and plants and so enter food chains, thereby risking exposure of populations to potentially hazardous levels of radioactivity. Inevitably, in the absence of a working disposal technology, an increase in the quantity of radioactive waste in storage will mean an undesirable increase in the risk of radioactivity escaping into the biosphere. Should the levels of escaping material reach significant proportions, an increase in the overall cancer death rate could well result.

2. In fact, if one assumes that a lung burden of ten micrograms of plutonium-239 is required to cause lung cancer (as opposed to the three micrograms found by Bair and Thompson (1974: 720) to cause cancer in the lungs of beagle dogs), the quantity of this isotope contained in the annual fuel offload of a 1000 MWe reactor would be the equivalent of about fifteen billion cancer-causing doses.

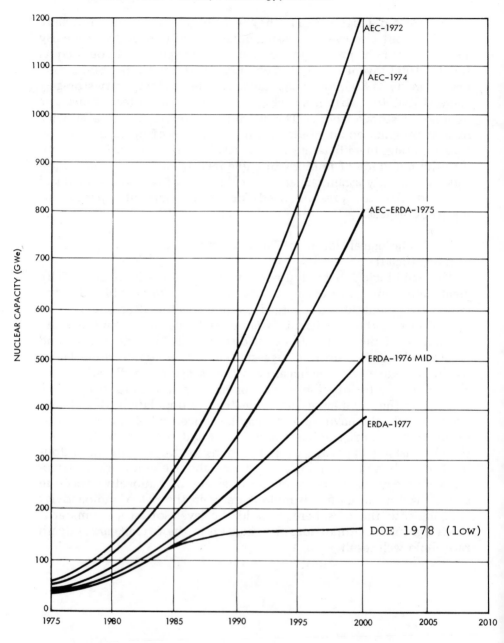

Figure 6-1. A Comparison of Recent Nuclear Power Growth Estimates.

Source: The DOE estimate is taken from IRG (1978). JPL (1977:4–8).

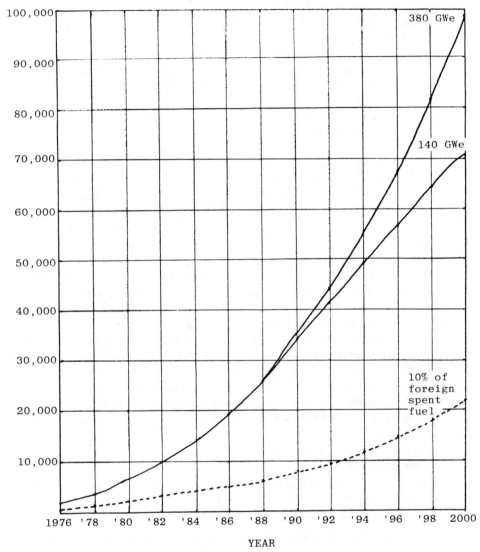

Figure 6–2. Cumulative Amount of Spent Reactor Fuel that will be in Storage Between the Years 1976 and 2000. *(1 metric ton = 2,200 pounds; the volume of 1 metric ton of reactor fuel is approximately 13.1 cubic feet.)*

Source: IRG (1978).

Although Various Waste Management and Disposal Technologies Have Been Proposed, None Has Yet Been Demonstrated to be Satisfactory

We have described proposed waste management and disposal technologies at some length. The customary approach involves solidification of liquid waste into glass blocks followed by encapsulation in metal canisters and emplacement of the canisters in geologic formations or, perhaps, seabed sediments. Ultimately, transmutation of some actinides or possibly disposal into solar orbit might become practical. Other disposal options are too costly, too risky, or too impractical. But even the most advanced technology—geologic isolation—is still no more than a promising concept. Large-scale vitrification of liquid waste is still years from implementation in the United States, and serious questions being raised about the suitability of the glass matrix suggest that it could turn out to be a poor choice. No one has actually constructed and tested a waste canister that will last for more than a few decades, even though many proposals are predicated on canister survival for centuries. Aside from some brief and generally inconclusive experiments in the 1960s, no one has actually placed a canister of radioactive waste in salt or granite or seabed sediment. Undoubtedly, the engineering capability to place a canister of waste in a geologic formation does now exist; the trick, in the words of Senator John Glenn, is whether we can ensure that the wastes won't "come bubbling up 100 miles away 50 years later" (*Nucleonics Week*, 19791: 6). The many uncertainties in existing data suggest that we cannot yet ensure that such a mishap or some quite different one will not occur.

Radioactive Waste Has Been Badly Managed in the Past

Recently, Carroll Wilson, the first general manager of the Atomic Energy Commission, writing about the early days of radioactive waste management, stated:

> Chemists and chemical engineers were not interested in dealing with waste. It was not glamorous; there were no careers; it was messy; nobody got brownie points for caring about nuclear waste. The Atomic Energy Commission neglected the problem. . . . The central point is that there was no real interest or profit in dealing with the back end of the fuel cycle. (1979: 15)

As a result of this quasi-official disinterest, the waste management program was severely compromised, with leaks of waste and instances of inexcusable incompetence becoming the rule rather than the exception. Even where storage measures seemed to be clearly inadequate from an early date, as in the case of the Hanford tanks, the AEC chose to disregard or downplay the scope and significance of the problem. As with many other nuclear-related matters under its jurisdiction—the hazards of weapons-testing fallout and reactor safety are two such examples—the AEC chose to ignore, mislead, or deceive the public about radioactive wastes.

We still face the consequences of this official mismanagement today. Leaks of radioactive wastes have become commonplace, and uncontrolled old radiation dumps, their locations lost, are being rediscovered with alarming regularity. A more serious casualty of the AEC's failures has been government credibility on the waste issue. The perhaps naive faith in the infallibility of the federal government's actions that existed in the 1950s has been replaced by a public cynicism and distrust that will constitute a continuing burden on future efforts to safely dispose of radioactive wastes. This is the legacy of the Atomic Energy Commission.

The Current Federal Program is Technically Superior to Previous Programs but Nonetheless Suffers from Several Serious Deficiencies

As we have made clear, the management and disposal of radioactive wastes has come to be recognized from both a health and economic viewpoint as a problem most critical to the future of the nuclear power program. This concern appears to be evenly spread over many sectors of our society. Nonetheless, the common concern has not produced a convergence of action. Various interest groups see the waste problem as a means to an end: the government, to make its energy policy more attractive; the nuclear industry, to achieve economic viability; and the environmental movement, to halt nuclear power. The desire to eliminate this environmental pollutant is frequently secondary. One result of these conflicting interests has been development of a program of "technical fixes" geared to getting the wastes out of sight and out of the mind of the public as soon as possible. The Waste Isolation Pilot Plant is a case in point.

If the WIPP project proceeds as planned, a "demonstration" of spent fuel disposal will commence around 1986, thereby, it is hoped, "solving" the disposal problem. Yet, all preliminary evidence suggests otherwise: the WIPP site appears to be technically defective; a convincing technical case for successful disposal cannot be made in five or even ten years; the superiority of salt as a disposal medium, particularly with regard to resource availability, must be questioned;

and an unsuccessful demonstration at WIPP might not only preclude recovery of the spent fuel, but could also cripple future disposal programs. The political commitment to WIPP is great—so great, apparently, as to prevent the Department of Energy from gracefully cancelling the project. Prudently discontinuing WIPP would mean at worst a postponement of demonstrated fuel disposal for perhaps five or ten years, no great price to pay in order to increase the chances for a successful program. Furthermore, other portions of the program appear to be better conceived from a technical point of view and could well provide the basis for construction of a successful repository in the 1990s.

But what about the intervening period? Government actions during this time must be of concern, for in the absence of a solution to the waste problem, temporary storage facilities will proliferate. A continued dependence upon these temporary expedients will ensure a continuing and increasing number of leaks of waste into the environment. Continuation of the domestic nuclear power program in the face of a prolonged failure to resolve the waste issue will surely be seen as irresponsible. The option of a limitation on the further generation of radioactive wastes, once viewed as the cry of an extremist minority, must now be considered a serious possibility. Indeed, such a limitation was suggested in 1977 by J. Gustave Speth, a member of the President's Council on Environmental Quality (*Nucleonics Week*, 1977a: 1), and in 1978 by Commissioner Peter Bradford of the Nuclear Regulatory Commission (*Nucleonics Week*, 1978b: 9).[3]

As this review makes clear, the radioactive waste problem is not one simply amenable to technical fixes. A solution will require careful consideration of not only technical, but also societal and political requirements. We discuss these requirements in the following section.

REQUIREMENTS FOR A SUCCESSFUL PROGRAM

A successful radioactive waste management and disposal program must address three types of requirements. The program must be technically feasible, it must be politically palatable, and it must be societally acceptable. It is our view that, of these three areas, only the matter of technical feasibility has been addressed in the past in any depth, and even here the treatment has been inadequate. Conse-

---

3. Furthermore, according to DOE, as of early 1979, a total of thirty states had enacted legislation or were planning to enact legislation that in some way would constrain waste disposal within their borders.

quently, no program of technical promise has ever been successfully implemented. Unless all three requirements are given at least equal consideration in program development, lack of convincing success in the future is almost a certainty. We will briefly address each of these requirements in turn.

Technical Feasibility

The technical prospects for developing a satisfactory means of radioactive waste disposal are, as much of our study illustrates, difficult to assess with total confidence, more because of significant unknowns than because of fundamental technical obstacles clearly in view. It is our judgement that technical problems can be largely overcome by investigations leading to judicious choice of disposal medium and site selection (of which we have more to say below), waste packaging and emplacement, and repository design and that none of these matters represents a fundamental technical obstacle. We believe that the necessary technology can be developed, surely for the shorter lived fission product waste, and that, at least in principle, the necessary degree of confidence in the technology can be achieved. Transuranic materials, with their long-lived radionuclides, and the longer lived fission products present greater problems, and although the required degree of confidence can probably be achieved for these wastes too, further research is required to establish this conclusively.

We cannot prejudge which avenues of study, which disposal technology, which waste form, which site will ultimately prove the best. And we are unable to make predictions concerning timetables for success. On these questions the jury is still out. One writer, assessing the view of earth scientists familiar with the waste disposal problem, has said: "Although these scientists continue to find the concept of geologic disposal attractive intuitively, some are stating explicitly that the scientific feasibility of the concept remains to be established" (Carter, 1978a: 1135). Nonetheless, geologic disposal is, in our view, a most fruitful, perhaps the only fruitful, approach to radioactive waste disposal available to us at the present time. This approach encompasses many technical possibilities and should form the centerpiece of the technical effort.

Too, the program must be broadly based. The problems and weaknesses of the current program lie not so much in the lack of technical possibilities as in the failure of the agencies responsible for the problem to implement a comprehensive program that includes careful investigation of all reasonable geologic alternatives. It is surely unwise to believe that single-minded pursuit of one solution, such as salt disposal, driven by political pressures and based largely on the notion of a "technical fix," can solve the problem. What such an approach disregards is that focus on a single path, and this one in particular, can subsequently cause great difficulties, if the technology proves unsuccessful, by foreclosing available alternatives. Given the current emphasis on salt, a mineral whose suitability as a disposal medium is still in doubt, the very real possibility exists that by 1995 or later, a solution to the waste problem may still not exist. Flexibility in the program is therefore essential.

Political Palatability

In order to identify, develop, and implement a satisfactory disposal technology, this country must carry through a program of research, development, and demonstration of significant challenge. The necessary foundation and basis for such a program requires the creation of an administrative and organizational structure within the federal government capable of providing the competence, perception, and authority that the resolution of a problem of this magnitude and importance will require. Unfortunately, this will have to be carried out in the climate of doubt and suspicion fostered by past weaknesses both technical and institutional and in the shadow of a consistent failure to fulfill promises and commitments to solve the problem. Indeed, this lack of confidence has already severely crippled the program. Thus, an additional requirement is the need for the program, as well as for the solution, to gain the confidence of a wary, if not antagonistic, public.

Societal Acceptability

President Carter has stated: "The waste generated by nuclear power *must* be managed so as to protect current and future generations" (Speth, 1979: 4). This must be the foremost criterion for any waste management program. This is the minimum demanded by society. Yet, it is clear that we cannot give a total guarantee that a waste repository will never be breached or that no person or persons of future generations will ever be harmed. Requirements to this end would block necessary moves and hinder needed progress because they are unrealistic and unachievable in the real world. What society can, and should, insist on is that the risks, which, we believe, can be made sufficiently small and can be bounded (so that, at worst, very few people would be hurt), are made very small. Given adequate time and sufficient funds to sustain a competent program, we believe that this goal can be accomplished to the satisfaction of the bulk of technically competent observers and critics and, most importantly, to the public at large.

Our recommendations, both institutional and technical, are directed toward avoiding the consequences of bureaucratic and technical inflexibility and incompetence. These proposals, presented below, reflect the belief that fundamental changes must be made in the structure and direction of the program if the mistakes of the past are not to be repeated. We address here only major requirements of a successful program: the need for unbiased management of and extensive public participation in program formulation and the need for a flexible technical program that will minimize the consequences of specific failures. We believe that the current program does not meet these needs and that, as a result, it will succeed badly, if it succeeds at all. Implementation of our recommendations would, we believe, be an important and useful beginning in dealing with the shortcomings of the current domestic radioactive waste management program and would go far in assuring the program's ultimate success.

RECOMMENDATIONS

Meeting Technical Requirements

Establish a Broadly Based Comprehensive Research and Development Program. Theoretical studies, while useful, may only add incrementally to existing knowledge. A program based upon extensive in situ research should be implemented as soon as possible. The program should be focused on disposal of wastes in several geologic media and performed only at sites deemed potentially usable for full-scale, high level waste repositories. Because the suitability of a given site is so highly dependent upon the particular characteristics of the site, decisions to develop a site should be willingly abandoned if serious defects become apparent at any stage of repository development. Such a program must be placed in competent hands, completely thorough in its investigation and implementation, and adequately funded so as to preclude attempts to take shortcuts in the process.

Repository Development and Construction Should Progress Through a Step-by-step Process. A program that attempts to jump directly from theoretical studies to demonstrations risks not only failure but also future public distrust and prejudice. Thus, a program that proceeds in incremental steps should be established.

Potential areas suitable for repository siting in all parts of the United States should be surveyed. Once sites are selected, experimental testing in mined vaults should be initiated, using a relatively small number of spent fuel elements. These vaults could be given tentative Nuclear Regulatory Commission licensing on the basis of applicable site suitability criteria. If these initial tests are deemed successful, the vault could be expanded into a licensed demonstration facility with a larger number of spent fuel elements. Finally, after a certain period of successful demonstration, the facility could be expanded into a full-scale, fully licensed, high level waste repository. Although not all sites would ultimately be converted into full-scale repositories, any such facility could only be developed as a part of this research and development sequence.

Abandon the Waste Isolation Pilot Plant Project as it is Currently Conceived. The WIPP site is technically unsuitable as a location for a radioactive waste repository. It may be penetrated by a breccia pipe. It is located fairly close to a circulating groundwater system; several faults run through it; it may be subject to low level earth tremors. It appears to lie over an anticline (deemed sufficient reason to abandon an earlier WIPP site), and perhaps most significantly, it is intimately associated with various natural resources that are likely to be exploited at some time in the future. These technical problems strongly suggest that the WIPP site should be abandoned.

WIPP is also suspect on political grounds. It is a project intended to demonstrate that nuclear waste disposal is feasible, in the absence of a detailed preliminary research program, thereby "proving" that the problem is solved. WIPP is not likely to meet the as yet undeveloped suitability criteria for repository siting. Unless plans are changed once again, WIPP is not intended to function as a full-scale repository, and the site-specific information gathered at the chosen location may be only of marginal use elsewhere. Finally, WIPP is officially intended only as a disposal site for alpha-contaminated military wastes that could also be buried in a full-scale, high level waste repository. For these reasons, we conclude that WIPP is intended to fulfill a political role—offering a potentially flawed "solution" to the radioactive waste problem—and is therefore not needed. The political promise of "proof" is too great a price to pay for the hasty implementation of a premature solution.

Ensure Full Investigation of Promising Alternatives. If, in the future, alternative disposal technologies appear more promising, they should be thoroughly researched. For the present, no nongeologic technology that appears environmentally, technologically, and economically feasible should be prematurely discarded.

Meeting Political Requirements

Establish an Independent Radioactive Waste Management Authority. Because a definitive solution to the waste problem is considered essential by nuclear proponents to the further expansion of nuclear power, there is, as we have noted, great pressure on the Department of Energy to implement a near-term solution. As with its predecessor agencies, DOE is caught in a similar, although not identical, conflict of responsibilities—promotion of nuclear power and disposal of nuclear wastes. Even though the Nuclear Regulatory Commission ostensibly regulates waste disposal, we believe the conflict to be a serious one, and we do not believe that DOE can, nor should be asked, to resolve it. The department cannot do so successfully. We believe that the management required for a successful radioactive waste disposal program cannot be situated within any existing agency of the federal government.

Thus, we recommend the establishment of a new authority, independent of the Department of Energy, that would assume control of the radioactive waste management program. The authority would develop technical research and development programs, determine waste management policy, and supervise the day-to-day management of the scientific and engineering aspects of the program. The technical and administrative staff of the authority should be composed of the most competent independent scientists, engineers, and administrators, and the director of the authority should be a distinguished individual with no previous ties to the nation's nuclear energy program who can restore a much-needed credibility to the radioactive waste management program. The result of establishing such an agency would not be the uncoupling of the futures of nuclear power and nuclear waste, but rather the elimination of the political and economic pressures that in themselves pose formidable obstacles to finding and developing a safe, sound solution to the radioactive waste problem.

Meeting Societal Requirements

Ensure Maximum Public Participation in Policy Decisionmaking and Waste Facility Siting in Order to Achieve a Societal Consensus on the Acceptability of the Program. No radioactive waste management program will ever prove acceptable or achieve success without public support. We recommend that the evolving program include an intensive public information program intended to inform the public about all aspects of the waste management program, including publicly financed dissemination of views that may run contrary to the views and findings of the program managers. We further recommend that the program include the public and the states in the decision-making and site suitability review processes. Citizen panels, public meetings with waste authority managers and scientists, annual reports to the public and Congress on the progress of the program, and adequate funding—to be made available by a direct tax on nuclear-generated electricity—for independent environmental, social, economic, and technical evaluations of waste management programs and facilities are all elements necessary to the ultimate success of the program. We believe that a waste management program that maximizes public education and involvement can and will minimize public objections to the goals of the program thus greatly enhancing the program's chances of success.

✳

Appendixes

Appendix A

Radionuclides in
Radioactive Waste

At the time of reactor shutdown, spent uranium reactor fuel contains some 461 fission product and 82 transuranic nuclides. Not all are radioactive, and of those that are, many decay to innocuous levels within minutes or hours. This appendix contains information about those radionuclides present in large quantity or with long half-lives.

Table A—1—This table lists radionuclides with half-lives greater than one day or that contribute more than 0.1 percent of fission product or transuranic activity during the first ten years after discharge, or that have very long half-lives.

Table A—2—This table lists specific activities of selected radionuclides in curies per gram and grams per curie.

Figure A—1—This graph shows the relationship between the activity and half-life of a radionuclide.

Tables A—3 through A—7—These tables list radionuclides according to half-lives and radionuclide inventories as a function of time. The tables also list the critical organs for exposure to the nuclides.

Table A–1.   Radionuclides in Spent Reactor Fuel.

Number	Radionuclide	Half-Life^a	Radiation
1	Tritium (H-3)	12.3 yr	beta
2	Cobalt–58	71.0 d	beta, gamma
3	Cobalt–60	5.26 yr	beta, gamma
4	Krypton–85	10.8 yr	beta, gamma
5	Rubidium–86	18.7 d	beta, gamma
6	Strontium–89	52.1 d	beta
7	Strontium–90	28.1 yr	beta
8	Yttrium–90	2.67 d	beta, gamma
9	Yttrium–91	59.0 d	beta, gamma
10	Zirconium–93	900,000 yr	beta, (gamma)^b
11	Zirconium–95	65.2 d	beta, gamma
12	Niobium–95	35.0 d	beta, gamma
13	Molybdenum–99	2.8 d	beta, *gamma*^{c}
14	Technetium–99	210,000 yr	beta
15	Ruthenium–103	39.5 d	beta, *gamma*
16	Ruthenium–106	366 d	beta, (gamma)
17	Rhodium–105	1.50 d	beta, gamma
18	Palladium–107	7,000,000 yr	beta
19	Tellurium–125m	58 d	beta, gamma
20	Tellurium–127m	109 d	beta, gamma
21	Tellurium–131m	1.25 d	beta, gamma
22	Tellurium–132	3.25 d	beta, gamma
23	Antimony–125	2.7 yr	beta, gamma
24	Antimony–127	3.88 d	beta, gamma
25	Iodine–129	17,000,000 yr	beta, gamma
26	Iodine–131	8.05 d	beta, gamma
27	Xenon–133	5.28 d	beta, gamma
28	Cesium–134	2.05 yr	beta, gamma
29	Cesium–135	2,000,000 yr	beta
30	Cesium–137	30.0 yr	beta, (gamma)
31	Barium–140	12.8 d	beta, gamma
32	Lanthanum–140	1.67 d	beta, gamma
33	Cerium–141	32.3 d	beta, gamma
34	Cerium–143	1.38 d	beta, gamma
35	Cerium–144	285 d	beta, *gamma*
36	Praseodymium–143	13.7 d	beta
37	Praseodymium–147	2.62 yr	beta, gamma
38	Neodymium–147	11.1 d	beta, gamma
39	Samarium–151	87 yr	beta, gamma
40	Europium–154	16 yr	beta, gamma
41	Europium–155	1.8 yr	beta, gamma
42	Protactinium–233	27 d	beta, gamma
43	Uranium–234	247,000 yr	alpha, gamma
44	Uranium–235	710,000,000 yr	alpha, gamma
45	Uranium–236	24,000,000 yr	alpha, gamma
46	Uranium–238	4,510,000,000 yr	alpha, gamma
47	Neptunium–237	2,100,000 yr	alpha, gamma
48	Neptunium–239	2.35 d	beta, gamma

Table A−1. continued

Number	Radionuclide	Half-Life[a]		Radiation
49	Plutonium-238	86	yr	alpha, gamma
50	Plutonium-239	24,400	yr	alpha, gamma
51	Plutonium-240	6,580	yr	alpha, gamma
52	Plutonium-241	13.2	yr	beta, gamma
53	Plutonium-242	380,000	yr	alpha, gamma
54	Americium-241	458	yr	alpha, gamma
55	Americium-242m	152	yr	alpha, gamma
56	Americium-243	7,950	yr	alpha, gamma
57	Curium-244	17.6	yr	alpha, gamma
58	Curium-245	9,300	yr	alpha, gamma

In addition to these radionuclides, after several thousand years other radionuclides, produced by the decay of transuranic elements (so-called "daughters"), build up in the spent fuel. After about 1 million years, the daughters reach secular equilibrium and remain fairly constant in terms of radioactivity. These daughters include Uranium-233, Thorium-230, Thorium-229, Actinium-229, Radium-226, Radium-225, Francium-221, Radon-222, Astatine-217, Polonium-218, Polonium-214, Polonium-213, Bismuth-214, Bismuth-210, Lead-210, and Lead-209. (There are also others.) The total radioactivity of these daughters is never much greater than about 15 curies.

[a] d = day; yr = year.
[b] (gamma) indicates radiation from unlisted short-lived daughter.
[c] *gamma* indicates radiation from radionuclide and unlisted short-lived daughter.

Sources: Hollocher (1975:226); JPL (1977:A-8); Bell (1973: table A-IV-1); NRC (1975: appendix VI, p. 3-3).

Table A–2.   Specific Activities of Important Nuclides in Radioactive Waste.

Nuclide	Half-life (years)	Grams/Curie	Curies/Gram
Fission products			
H-3	12.3	0.0001	3,600
Co-60[a]	5.26	0.0009	1,140
Sr-90	28.1	0.0071	140
Zr-93	900,000	390	0.0026
Tc-99	210,000	58	0.017
Ru-106	1.0	0.0004	2,500
I-129	17,000,000	6,100	0.00016
Cs-134	2.05	0.00074	1,350
Cs-135	2,000,000	1,300	0.00077
Cs-137	30.0	0.011	91
Ce-144	0.78	0.00032	3,120
Pm-147	2.62	0.0011	910
Sm-151	87	0.036	28
Eu-154	16	0.071	141
Eu-155	1.8	0.00078	1,280
Actinides			
U-234	247,000	160	0.0062
U-235	710,000,000	460,000	0.000002
U-236	24,000,000	20,000	0.00005
U-238	4,510,000,000	3,000,000	0.0000003
Np-237	2,100,000	1,400	0.0007
Pu-238	86	0.057	17.5
Pu-239	24,400	16	0.062
Pu-240	6,600	4.4	0.23
Pu-241	13.2	0.0088	114
Pu-242	380,000	260	0.0038
Am-241	458	0.031	32.2
Am-242m	150	0.1	10
Am-243	7,950	5.4	0.18
Cm-243	32	0.022	45.4
Cm-244	17.6	0.012	83.3

[a]Co-60 is not a fission product but is formed by neutron activation in the fuel cladding.

Sources: NAS (1978:216); EPA (1978b: table A-II.8).

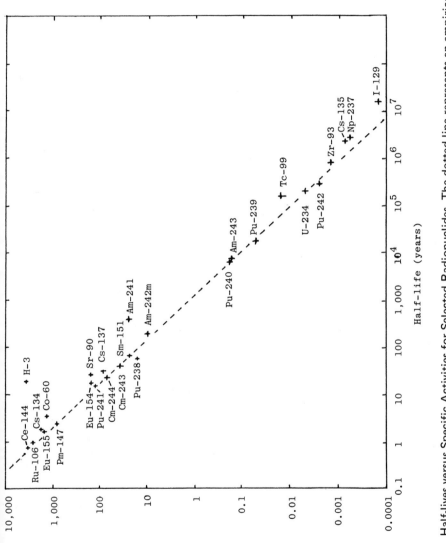

Figure A–1. Half-lives versus Specific Activities for Selected Radionuclides. The dotted line represents an empirical fit to the data points.

Source: Prepared by author.

Table A−3.   Radionuclides in Spent Reactor Fuel with Half-lives Shorter than Ten Years.

Radionuclide	Half-life	Radiation	Critical Organ	Radioactivity (in curies) per Metric Ton						
				$t = 0^a$	100 d	1 yr	5 yr	10 yr	25 yr	50 yr
Ce-144/Pr-144[b]	285 d	beta, gamma	bone, GI tract	$2.1M^c$	1.6M	864k	25k	300	—	—
Ru-106/Rh-106	1 yr	beta, gamma	GI tract	1.1M	910k	550k	34k	1,100	—	—
Eu-155	1.8 yr	beta, gamma	GI tract, kidney	7,500	6,750	5,100	1,100	160	—	—
Cs-134	2.05 yr	beta, gamma	whole body	240k	219k	171k	44k	8,300	51	—
Pm-147	2.62 yr	beta, gamma	GI tract, bone	110k	102k	84k	29k	7,900	148	—
Sb-125/Te-125m	2.7 yr	beta, gamma	kidney	12,750	11,900	9,860	3,530	980	21	—
Co-60[d]	5.26 yr	beta, gamma	GI tract	8,930	8,610	7,830	4,620	2,390	330	12

[a] Radioactive inventory of fuel at time of discharge from reactor. Fuel burnup of approximately 33,000 MWd/ton. Numbers are approximate. d = day; yr = year.

[b] The second isotope in each set is a short half-life daughter of the first isotope. The half-lives are as follows: Pr-144, 17.3 min.; Rh-106, 130 min.; Te-125m, 58 d.

[c] M = megacuries (millions of curies); k = kilocuries (thousands of curies).

[d] Cobalt-60 is present in the fuel cladding.

Sources: Hollocher (1975:225); JPL (1977: A-8); Bell (1973: table A-IV-1).

Table A–4. Radionuclides in Spent Reactor Fuel with Half-lives Between Ten and Twenty Years.

Radionuclide	Half-life	Radiation	Critical Organ	$t = 0^a$	Radioactivity (in curies) per Metric Ton					
					5 yr	10 yr	25 yr	50 yr	100 yr	500 yr
Kr–85	10.8 yr	beta, gamma	lung, whole body	11.4k[b]	8,275	6,000	2,300	460	18	—
H–3 (tritium)	12.3 yr	beta	whole body	750	565	425	180	45	3	—
Pu–241	13.2 yr	beta, gamma	liver, lymph nodes	125k	96k	74k	34k	9,050	655	—
Eu–154	16 yr	beta, gamma	GI tract, kidney	7,800	6,280	5,060	2,640	890	102	—
Cm–244	17.6 yr	alpha, gamma	bone	1,950	1,600	1,320	730	270	38	—

[a] Radioactive inventory of fuel at time of discharge from reactor. Fuel burnup of approximately 33,000 MWd/ton. Numbers are approximate. d = day; y = year.
[b] k = kilocuries (thousands of curies).
Sources: Hollocher (1975:225); JPL (1977:A-8); Bell (1973: table A-IV-1).

Table A-5.   Radionuclides in Spent Reactor Fuel with Half-lives Between Twenty and One Thousand Years.

Radionuclide	Half-life	Radiation	Critical Organ	Radioactivity (in curies) per Metric Ton						
				$t = 0^a$	5 yr	10 yr	50 yr	100 yr	500 yr	1,000 yr
Sr-90/Y-90[b]	28.1 yr	beta, gamma	bone	155k[c]	137k	121k	45.2k	13.2k	1	—
Cs-137/Ba-137m	30.0 yr	beta, gamma	whole body	208k	185k	165k	65.5k	20.6k	2	—
Pu-238	86 yr	alpha, gamma	liver, lymph nodes	2,300	2,210	2,120	1,540	1,030	41	—
Sm-151	87 yr	beta, gamma	GI tract, bone	1,260	1,210	1,160	850	575	23	—
Am-241[d]	458 yr	alpha, gamma	kidney, bone	2,030	2,900	3,730	3,810	3,860	2,060	924

[a] Radioactive inventory of fuel at time of discharge from reactor. Fuel burnup of approximately 33,000 MWd/ton. Numbers are approximate.   d = day; yr = year.

[b] The second isotope in each set is a short half-life daughter of the first isotope. The half-lives are as follows: Y-90, 64 hr.; Ba-137m, 2.6 min.

[c] k = kilocuries (thousands of curies).

[d] Am-241 inventory increases as a result of decay of Pu-241.

Sources: Hollocher (1975:225); JPL (1977:A-8); Bell (1973: table A-IV-1).

Table A–6. Radionuclides in Spent Reactor Fuel with Half-lives Between 1,000 and 100,000 Years.

Radionuclide	Half-life	Radiation	Critical Organ	Radioactivity (in curies) per Metric Ton					
				$t = 0^a$	100 yr	1,000 yr	10,000 yr	100,000 yr	1,000,000 yr
Pu–240[b]	6,580 yr	alpha, gamma	bone, red marrow	479	485	435	170	—	—
Am–243	7,950 yr	alpha, gamma	kidney, bone	17.0	17.0	15.5	7.1	—	—
Pu–239[c]	24,400 yr	alpha, gamma	bone, red marrow	323	323	314	239	191	—

[a] Radioactive inventory of fuel at time of discharge from reactor. Fuel burnup of approximately 33,000 MWd/ton. Numbers are approximate. yr = year.

[b] Pu–240 inventory increases as a result of decay of Cm–244.

[c] Pu–239 inventory increases slightly as a result of decay of Am–243.

Sources: Hollocher (1975:225); JPL (1977:A–8); Bell (1973:table A–IV–1).

Table A–7. Radionuclides in Spent Reactor Fuel with Half-lives Greater than 100,000 Years.

Radionuclide	Half-life (thousands of years)	Radiation	Critical Organ	Radioactivity (in curies) per Metric Ton				
				$t = 0^a$	10,000 yr	100,000 yr	1,000,000 yr	10,000,000 yr
Tc-99	210	beta	GI tract	15	15	11	0.6	—
U-234[b]	247	alpha, gamma	bone, GI tract	1.1	1.9	1.5	0.41	—
Pu-242	380	alpha, gamma	bone	1.8	1.8	1.5	0.3	—
Zr-93	900	beta, gamma	GI tract	1.9	1.9	1.8	1.7	—
Cs-135	2,000	beta	whole body	0.2	0.2	0.2	0.18	0.01
Np-237[c]	2,100	alpha, gamma	bone	0.3	1.2	1.2	0.9	0.03
Pd-107	7,000	beta	GI tract, kidney	0.013	0.013	0.013	0.012	0.005
I-129	17,000	beta, gamma	thyroid	0.045	0.045	0.045	0.043	0.030
U-236[d]	24,000	alpha, gamma	GI tract, kidney	0.26	0.35	0.40	0.39	0.29
U-235[e]	710,000	alpha, gamma	GI tract, kidney	0.02	0.02	0.02	0.03	0.03
U-238[f]	4,510,000	alpha, gamma	GI tract, kidney	0.3	0.3	0.3	0.3	0.3

[a] Radioactive inventory of fuel at time of discharge from reactor. Fuel burnup of approximately 33,000 MWd/ton. Numbers are approximate. yr = year.

[b] U-234 inventory increases as a result of decay of Pu-238. U-234 decays to Th-230.

[c] Np-237 inventory increases as a result of decay of Pu-241 and Am-241. Np-237 decays to U-233.

[d] U-236 inventory increases as a result of decay of Pu-240. U-236 decays to Th-232.

[e] U-235 inventory increases as a result of decay of Pu-239 and Am-243.

[f] U-238 inventory increases slightly as a result of decay of Pu-242. Other long-lived isotopes present in reactor fuel are: U-233, 160,000 years; Th-230, 77,000 years; Th-232, 14 billion years. None of these isotopes contributes appreciably to radioactivity.

Sources: Hollocher (1975:225); JPL (1977:A–8); Bell (1973: table A–IV–1).

❋ *Appendix B*

Health Risks Associated with Failed Radioactive Waste Management Facilities

We have performed four simple risk analyses in order to evaluate and compare the consequences and risks to an exposed population as a result of failed waste storage and disposal facilities. These analyses are based upon Cohen's (1977a) dose equations but assume a higher dose-risk relationship than used in his articles. These analyses are not meant to be either definitive or all-inclusive. They do, however, allow a degree of relative comparison or risks resulting from different events.

The following four cases were analyzed:

Case 1:   Aging high level liquid waste storage tanks leak 400,000 gallons of waste containing 250,000 curies of strontium-90 into the ground. After a one hundred year delay, 10 percent of the Sr-90 enters a river located one mile away. The river, with an annual flow of 500 million cubic meters, supplies drinking water to a population of 500,000.

Case 2:   One hundred years after the sealing of a bedded salt repository containing 100,000 metric tons of spent fuel, an undetected fault becomes active and allows groundwater to enter the repository and carry away radioactive waste. During a one year period, 7.5 metric tons of spent fuel are leached away.[1] One hundred years later, 10 percent of the Sr-90 in the leached fuel reaches the river described above.

1. The annual leach rate used here is one-half that assumed to occur after facility disruption (0.015 percent of capacity per year) in the Arthur D. Little, Inc. study (EPA, 1978a) referred to in the text.

Case 3:   Ten thousand years after the sealing of a bedded salt repository containing 100,000 metric tons of spent fuel,
groundwater enters the facility and begins to carry away
radioactive waste. During a one year period, 7.5 metric
tons of spent fuel are leached away. One thousand years
later, 100 percent of the technetium-99 in the leached
fuel reaches the river described above.

Case 4:   As a worst case analysis, we assume that 250,000 curies of
Sr-90 (referred to in Case 1) enter the river described
above. The leak is discovered almost immediately, and as a
result, ingestion of the contaminated water is limited to
one day.

Throughout these analyses, the following assumptions have also
been made:

1. Each individual in the exposed population consumes 2.2 liters
   of contaminated water daily;
2. The contamination remains undetected for one year (except in
   Case 4);
3. Death from cancer due to exposure occurs during the thirty years
   following a latency period of fifteen years;
4. Each ten thousand person rem of exposure results in one to two
   fatal cancers.[2]

A typical calculation is performed as follows:

1. A quantity of radioactive material (Q_r) enters a volume of water
   (V_w). The concentration of radioactivity in the water (C_r) is:

$$C_r \text{ (curies/cubic meter)} = \frac{Q_r \text{ (curies)}}{V_w \text{ (cubic meters)}} .$$

2. The dose to an exposed individual (D_r) is calculated by dividing
   C_r (concentration of the radionuclide in water) by the maximum
   permissible concentration for occupational exposure (MPC_w) in
   water of the radionuclide. The occupational MPC_w is that concentration of a radionuclide (in curies per cubic meter) that, if

2. These are the dose-risk quantities generally assumed by the Environmental
Protection Agency (one cancer per 10,000 person rem) and used in a recent re-
port by the National Academy of Sciences (one cancer per 5,000 person rem)
(NAS, 1979: 12).

the water is ingested at a rate of 2.2 liters per day for a year, would give a dose commitment d_j to body organ j, where d_j = 30 rem for bone and thyroid and 15 rem for lungs, kidneys, gastrointestinal tract, and other body organs. The MPC_w for Sr-90 is 4 microcuries per cubic meter, and for Tc-99, 3 millicuries per cubic meter. D_r is thus given by:

$$D_r \text{ (rems)} = \frac{C_r \text{ (curies/cubic meter)}}{MPC_w \text{ (curies/cubic meter)}} \times d_j \text{ (rems)}.$$

3. The total risk to an individual is calculated. In these analyses the lifetime risk of cancer to a young adult is calculated assuming an absolute risk model with a latency period of fifteen years followed by a thirty year period of peak risk. Additional cancers may occur after this period but would be difficult to detect among all non-radiation-caused cancers. For children, the risk will be greater; for older adults, it will be smaller. We assume an average risk for the entire population. The total risk of developing a fatal cancer (R_f) as a result of radiation exposure is:

$$R_f = D_f \text{ (rems)} \times \text{risk/rem.}[3]$$

4. The risk is distributed over a population of 500,000. Thus, after fifteen years the annual fatality rate as a result of radiation-induced cancer (F_r) is:

$$F_r \text{ (deaths/year)} = \frac{500{,}000 \times R_f \text{ (risk/person)}}{30 \text{ years}}.$$

The current domestic cancer death rate is about 1,600 deaths per year per million population.

CASE 1: HIGH LEVEL LIQUID WASTE LEAKS FROM OLD STORAGE TANKS

What is the probability of such an event occurring? We know from experience that, with inept management, such an event is by no means implausible. As a basis for this calculation, we have used the total of the many leaks at Hanford Reservation: 422,000 gallons of

---

3. Risk/rem is simply the inverse of the number of fatal cancers expected per 10,000 person rem.

waste containing some 500,000 curies of radioactivity, of which approximately 50 percent is strontium-90. We assume that the leaks take place over a period that is short in relation to the time it takes the Sr-90 to reach the river. The liquid enters directly into the ground and begins to move toward the river, located one mile away.

We have considered only the hazard of Sr-90, because this radionuclide dominates the hazard of fission product waste during the first few centuries after production. The movement of Sr-90 through rock is typically retarded by a factor of one hundred due to ion exchange. We have here used a smaller retention factor and assume that it takes one hundred years for 10 percent of the Sr-90 to reach the river. This quantity enters the river during the one hundredth year.

The 250,000 curies of Sr-90 decay to approximately 21,250 curies by the one hundredth year. Of this, only 2,150 curies enter the river. The concentration of Sr-90 in the water is:

$$\frac{2{,}125 \text{ curies}}{500{,}000{,}000 \text{ cubic meters}} = 0.000004 \text{ curies/cubic meter.}$$

Over the course of one year, the radiation dose to an individual is:

$$\frac{0.000004 \text{ ci/cu. m.}}{0.000004 \text{ ci/cu. m.}} \times 30 \text{ rem/year} = 30 \text{ rem year.}$$

The risk to an individual is:

High risk: 30 rem $\times$ 0.0002 risk/rem = 0.006 risk of death;
Low risk: 30 rem $\times$ 0.0001 risk/rem = 0.003 risk of death.

The total death rate due to the one year exposure is:

High: 500,000 persons $\times$ 0.006 risk = 3000 deaths;
Low: 500,000 persons $\times$ 0.003 risk = 1500 deaths.

The annual death rate will lie between fifty and one hundred, an increase 6.25 to 12.5 percent over the normal eight hundred cancer deaths per year.[4]

---

4. We should note that in the case of the Hanford Reservation, the retention of Sr-90 in soil is quite large, so much so that little or no Sr-90 is likely to reach the Columbia River before most of the activity decays away.

CASE 2: STRONTIUM-90 ESCAPES
FROM A ONE HUNDRED YEAR OLD
SALT REPOSITORY

One hundred years after the sealing of a bedded salt repository containing 100,000 metric tons of spent reactor fuel, an undetected fault running through the facility becomes active and allows groundwater to enter. The groundwater begins to leach away the spent fuel, and during the course of one year, 7.5 metric tons of fuel are carried away. Again, Sr-90 dominates the hazard and will be the only radionuclide considered. Because the groundwater and waste are flowing through salt, ordinary ion retention processes are greatly reduced. We assume that the water takes one hundred years to reach the river, located some ten miles from the repository, and that 10 percent of the Sr-90 in the leached reactor fuel enters the river at this time.

One metric ton of one hundred year old spent reactor fuel contains about 13,200 curies of Sr-90. The total amount of Sr-90 leached away during one year is:

$$7.5 \text{ metric tons} \times 13,200 \text{ curies} = 99,000 \text{ curies.}$$

(We will round this up to an even 100,000 curies for convenience.) Over the following one hundred years, this quantity of Sr-90 decays to about 8,500 curies. Of this, 10 percent, or 850 curies, are discharged into the river during the one hundredth year. The concentration of Sr-90 in the water at this time is:

$$\frac{850 \text{ curies}}{500,000,000 \text{ cubic meters}} = 0.0000017 \text{ curies/cubic meter.}$$

The dose to an individual as a result of ingesting this water for one year is:

$$\frac{0.0000017 \text{ curies/cubic meter}}{0.000004 \text{ curies/cubic meter}} \times 30 \text{ rems} = 12.75 \text{ rems.}$$

The risk to an individual of developing a fatal cancer is:

High: 12.75 rems $\times$ 0.0002 risk/rem = 0.00255 risk;
Low: 12.75 rems $\times$ 0.0001 risk/rem = 0.00128 risk.

The total number of radiation-caused deaths as a result of the one year exposure is:

High: 500,000 persons $\times$ 0.00255 risk = 1275 deaths;
Low: 500,000 persons $\times$ 0.00128 risk =   638 deaths.

The annual death rate will range from twenty-one to forty-two, an increase of 2.6 to 5.2 percent over the normal cancer death rate.

Note that the one year cutoff for exposure is artificial; if the Sr-90 in the water were detected at an earlier time, cumulative exposure would be less; if detected later, a greater exposure would result. Also, the leaching of spent fuel would be a continuous process, and radioactive contamination of the water would continue for many years, if not decades or centuries.

CASE 3: TECHNETIUM-99 ESCAPES FROM A 10,000 YEAR OLD SALT REPOSITORY

We make the same assumptions as in Case 2, but instead consider the release of technetium-99, a fission product with a half-life of 210,000 years. Tc-99 is not retarded at all by ion retention processes, and therefore all the Tc-99 leached away by groundwater will eventually enter the river. We also assume an arbitrary delay in the time the Tc-99 takes to enter the river, which, in any case, is a small fraction of one half-life.

A metric ton of spent reactor fuel contains about fifteen curies of Tc-99. The 7.5 metric tons of fuel leached away each year contain about 110 curies of the radionuclide. The concentration of Tc-99 in the water after these 110 curies enter the river is:

$$\frac{110 \text{ curies}}{500,000,000 \text{ cubic meters}} = 0.00000022 \text{ curies/cubic meter.}$$

The dose to an individual as a result of a one year exposure is:

$$\frac{0.00000022 \text{ curies/cubic meter}}{0.003 \text{ curies/cubic meter}} \times 15 \text{ rems} = 0.0011 \text{ rem.}[5]$$

5. However, since exposure would occur over many years, the cumulative dose and risk would be greater.

The risk to an individual as a result of the one year exposure is:

High: 0.0011 rem $\times$ 0.0002 risk/rem = 2.2×10^{-7} risk;
Low: 0.0011 rem $\times$ 0.0001 risk/rem = 1.1×10^{-7} risk.

Thus, during the thirty year period of greatest risk, the number of cancer deaths resulting from a one year exposure to Tc–99 is:

High: 500,000 persons $\times$ 2.2×10^{-7} risk = 0.11 deaths;
Low: 500,000 persons $\times$ 1.1×10^{-7} risk = 0.055 deaths.

But this is really an annual death rate, because the Tc–99 continues to leak into the river every year. Over the first half-life of the radio-nuclide, some 11,500 to 23,100 cancer deaths might be expected to occur as a result of exposure to Tc–99. Nonetheless, this number would be lost in the 168 million cancer deaths arising from other causes.

CASE 4: WORST CASE ANALYSIS

The 400,000 gallons of liquid waste referred to in Case 1 contain approximately 250,000 curies of strontium–90. We assume that the entire quantity of Sr–90 enters the river over a one year period, but exposure to the radionuclide is limited to one day. The concentration of Sr–90 in the water is:

$$\frac{250,000 \text{ curies}}{500,000,000 \text{ cubic meters}} = 0.0005 \text{ curies/cubic meter.}$$

The dose to an individual as a result of ingesting the contaminated water for one day is:

$$\frac{0.0005 \text{ curies/cubic meter}}{0.000004 \text{ curies/cubic meter}} \times 30 \text{ rem/year} \times \frac{1}{365} \text{ year/day} = 10.3 \text{ rem/day.}$$

The risk to an individual of developing a fatal cancer as a result of this dose is:

High: 10.3 rems $\times$ 0.0002 risk/rem = 0.002 risk;
Low: 10.3 rems $\times$ 0.0001 risk/rem = 0.001 risk.

The total number of deaths over the thirty year period of greatest risk as a result of the one day exposure is:

> High: 500,000 persons × 0.002 risk = 1,000 deaths;
> Low: 500,000 persons × 0.001 risk = 500 deaths.

The annual death rate will range from seventeen to thirty-four, or an increase of about 2 to 4 percent over the normal eight hundred cancer deaths per year.

Appendix C

Foreign Programs in Radioactive Waste Management

To allow comparison with the waste management program of the United States, short descriptions of foreign waste management programs are presented below. Although some countries are farther advanced in particular waste management areas, such as vitrification, no country has progressed beyond the United States in the field of permanent waste disposal. Section 7 of the Bibliography contains an extensive list of relevant references.

UNITED KINGDOM

The U.K. has a reprocessing facility that is capable of handling about 1,000 metric tons per year. The reprocessing wastes are stored in liquid form in double-walled steel tanks. Eventually, this waste is supposed to be converted into borosilicate glass and buried in geologic formations. Current research points to two options: clay formations or crystalline rocks (Chapman, Gray, and Mather, 1978).

FRANCE

France has constructed or is in the process of constructing reprocessing facilities with an annual capacity of 3,200 metric tons. Liquid reprocessing wastes have been stored in engineered storage facilities until now. The French have developed a program for converting the liquid wastes to borosilicate glass. They are also assessing the suitability of salt as a medium for geologic disposal (DOE, 1978a).

CANADA

The Canadian waste management program generally tracks that of the United States. Presently, no fuel-reprocessing facilities for commercial fuel exist in Canada, and spent fuel is stored as in the U.S. Studies into the feasibility of geologic disposal have been conducted, concluding that igneous rock may be a suitable medium. As of yet, no actual disposal has taken place (DOE, 1978a).

JAPAN

At the present time, Japan has ten operating nuclear reactors and ten others under construction or in the planning stage. Current reprocessing capacity is 210 metric tons per year, although a 1,900 metric tons per year plant is planned for operation in the 1990s. Reprocessing wastes are stored as acid liquid in stainless steel tanks. Because Japan has no terminal disposal capability due to its small land area, the Japanese are very interested in activities in other countries related to geologic disposal. They are also interested in seabed disposal and the island disposal concept—that is, finding an uninhabited island and dumping the waste there (DOE, 1978a).

FEDERAL REPUBLIC OF GERMANY

The FRG has no reprocessing capacity for commercial fuel, but does have a commitment from France to reprocess all uncommitted German fuel through 1981. Germany plans to construct a fuel cycle center at Gorleben in Lower Saxony. At this site, fuel will be reprocessed and recycled, and waste will be buried in salt below the site. Some political problems may arise from this, however, because the Gorleben salt dome extends under the Elbe River into East Germany.[1] Currently, the Germans are disposing of low and intermediate level wastes in an abandoned salt mine at Asse (see Figures C−1 and C−2). This is a commercial program (Krugmann, 1978).

---

1. In the spring of 1979, the German federal government, under pressure from the public and the state of Lower Saxony (location of the Gorleben site), decided to postpone construction of the facility (see Section 7 of the Bibliography).

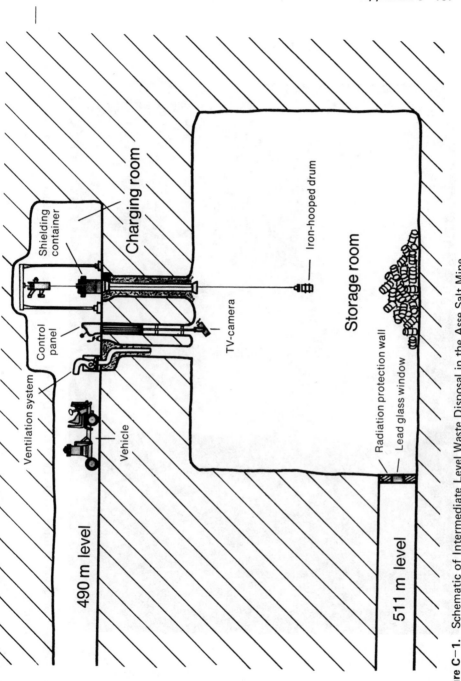

Figure C—1. Schematic of Intermediate Level Waste Disposal in the Asse Salt Mine.

Source: GSF (1973.:25).

Figure C–2. Photograph of Intermediate Level Waste Disposal Vault in the Asse Salt Mine, taken through a lead glass window.

Source: GSF (1973:27).

BELGIUM

Although Belgium has an active waste management program that is investigating, among other things, geologic disposal, it seems unlikely that any waste will be buried within the borders of the country as a result of its limited land area (DOE, 1978a).

SWEDEN

The Swedish government has a reactor-licensing policy similar to that of California. In order to comply with the "Nuclear Stipulation Law," as it is called, the Swedish power industry established the Nuclear Fuel Safety Project—also called the KBS Project—to develop and evaluate a method for the management of glassified liquid waste from reprocessing through storage in deep crystalline rock. Under the KBS plan, spent fuel will be reprocessed in France, the liquid wastes vitrified and placed in stainless steel canisters, returned to Sweden and allowed to cool for thirty years, with emplacement of the encapsulated wastes in a crystalline rock repository at a 500 meter depth no earlier than 2020. The storage holes and tunnels will be backfilled with a quartz sand-bentonite mixture that possesses good ion exchange characteristics and that upon contact with water expands to become almost totally impermeable. A review of the KBS Report by the California Commission (CERCDC, 1978b) concluded that many of the technical gaps troubling the U.S. geologic disposal program would also be encountered in the Swedish program. An assessment of the report written for the Swedish Energy Commission by Dr. J. Winchester (1978b) of the Florida State University Department of Oceanography presented similar conclusions. Sweden is also cooperating with the U.S. in tests being conducted in an abandoned iron mine to determine the response of granite to heating (*Industrial Research/Development*, 1978; OWI, 1978).

USSR

The Soviet Union has a very small reprocessing capacity at present, but is building a commercial reprocessing facility with an 1,800 MT/yr. capacity, to be operational in the early 1980s. Experiments with the vitrification of wastes into glass are being conducted, and studies of geologic isolation have been done. At present, surface storage for high level wastes is being emphasized.

One of the more intriguing stories concerning radioactive waste management has to do with a purported radioactive waste explosion that occurred in the Soviet Union. The primary source of information concerning this incident has been Zhores Medvedev, a Soviet emigré biologist currently residing in London. In an article in *New Scientist* (Medvedev, 1976), he wrote that a waste storage site in Kyshtym in the southern Ural Mountains exploded in 1957, killing hundreds, injuring thousands, and radioactively contaminating some 2,000 square kilometers (770 square miles) to very high levels. Although Medvedev assumed that the accident was known to Western scientists, such was not the case, and his report created a great deal of interest and no little controversy. Nonetheless, the possibility of such an accident was dismissed by most nuclear scientists in the United States and Great Britain. Medvedev, in an attempt to prove the truth of his report, began to research the incident and published two articles (1977a, 1977b) which confirmed that some sort of nuclear accident had indeed taken place. In late 1977, the CIA released documents which indicated that a nuclear accident—possibly associated with a reactor—had occurred in Kyshtym (White, 1977).

By 1978, Medvedev managed to arouse sufficient interest, hostility, and criticism to cause the formation of a study group on the Kyshtym accident at the Oak Ridge National Laboratory. The group published a preliminary report (Trabalka, et al., 1979) which stated that Medvedev erred about the source of the radioactive material and the size of the contaminated area. The study group concluded that the accident resulted from a chemical explosion in a nuclear fuel reprocessing plant and contaminated 65 square kilometers rather than the 2,000 reported by Medvedev. However, at least one member of the study group has acknowledged that Medvedev may have been correct about the source of the explosion (O'Toole, 1979). In any case, the amount of radioactivity released was enormous: up to one million curies of strontium-90 may have entered a lake near the site of the accident, resulting in a level of contamination about 1,000 times greater than that which would be caused by fallout from the explosion of a nuclear bomb (O'Toole, 1979). (See Section 7 of Bibliography for a listing of references about this accident.)

INDIA

India, with three small power reactors and five more under construction, has a 60 metric ton per year reprocessing plant and is planning to construct a 100 metric ton per year plant. India also possesses a plant capable of reprocessing fuel from an experimental reactor. Plutonium extracted from this fuel was used to construct the atomic

device detonated in 1974. India is investigating igneous rock formations and selected sedimentary deposits as repository sites (IRG,
1978).

OTHER COUNTRIES

Many other countries have nuclear programs, with reactors in operation, under construction, or on order. They do not have any reprocessing or disposal capacity, however. President Carter announced
in 1977 that the United States would be willing to accept limited
amounts of spent fuel from selected countries if such a move would
prevent construction of national reprocessing facilities with the attendant danger of nuclear proliferation. The Deutch Report estimates that by the year 2000, the United States may have up to
22,000 metric tons of foreign spent fuel in storage (DOE, 1978a:
141).

Glossary

Actinides—Series of elements beginning with actinium, no. 89, and continuing through lawrencium, no. 103. Includes uranium and all manmade transuranic elements.

Acute radiation exposure—Exposure to levels of radiation in excess of fifty rems over a period of twenty-four hours or less.

Alpha-contaminated waste—See *TRU waste.*

Alpha particle—A charged particle emitted from the nucleus of an atom, having a mass and charge equal in magnitude to a helium nucleus, easily stopped by several sheets of paper.

Anticline—A fold of rocks whose core contains strata of older rocks; it is convex upward.

Aquifer—A subsurface formation or geological unit containing sufficient saturated permeable material to yield significant quantities of water.

Atomic energy—Term relating to production of energy by nuclear fission; see *Fission.*

Away from reactor storage (AFR)—A storage facility for spent reactor fuel; see *Swimming pool.*

Background radiation—Radiation in the human environment from naturally occurring radioactive elements and cosmic radiation.

Basalt—An igneous rock of volcanic origin, usually fine-grained and black or dark gray.

Beta particle—Charged particle emitted from the nucleus of an atom, with a mass and charge equal in magnitude to that of an electron; stopped by a thin sheet of metal.

Biosphere—That part of the earth that contains life.

Borosilicate glass—A glass matrix into which high level liquid wastes are incorporated to 13 percent by weight fraction.

Breccia pipe—A geologic feature present in the vicinity of salt formations, it is indicative of dissolution of salt by flowing water. A pipelike formation that extends from the surface into the salt bed, filled with rubble.

Burnup—A measure of reactor fuel consumption, usually expressed in terms of the amount of energy produced per unit weight of fuel in the reactor—e.g., 33,000 megawatt days/metric ton, a typical burnup for uranium fuel in a light water reactor.

Calcine—A solid-grained form of high level reprocessed waste produced by a special drying process applied to liquid wastes.

Canister, waste—The outermost container into which glassified HLW or spent fuel is to be placed. Made of stainless steel or an inert alloy. Dimensions: approximately one and a half feet in width and ten feet high.

Cesium—Element 55. Cesium-137 is a radioactive fission product with a half-life of thirty years. Cesium is highly mobile in the biosphere.

Chain reaction—That process whereby fissioning of atoms continues without external influence. May be controlled or uncontrolled.

Chronic radiation exposure—Exposure to low levels of radiation over an extended period of time.

Cladding—Protective alloy shielding in which fissionable fuel is inserted. Cladding is relatively resistant to radiation and the physical and chemical conditions in a reactor core. May be stainless steel or some alloy such as zircalloy.

Concentration factor—That factor by which radioactivity is concentrated in an organism above ambient levels.

Crib—A covered trench into which low level liquid wastes are dumped. The wastes are assumed to percolate into the underlying soil layers.

Criticality—A set of physical conditions in which a nuclear chain reaction is self-sustaining. The principal variables influencing criticality include the amount of fissionable material (critical mass), its distribution in space, the presence or absence of neutron absorbers and of moderators (materials that slow down fast neutrons).

Critical mass—The minimum mass of fissionable material that, with a specified geometric arrangement and material composition, will self-sustain a fission chain reaction.

Curie—A measure of radioactivity equal to that of one gram of fresh radium per second. Approximately thirty-seven billion disintegrations per second.

Daughter—A nuclide formed by the radioactive decay of another nuclide, called the parent.

Decay—See *Radioactivity*.

Decay chain—Successive decays in a radioactive series, usually by alpha particle emission.

Decay heat—Heat produced by the decay of radioactive nuclides.

Decay product—A nuclide resulting from the radioactive disintegration of a radionuclide, formed either directly or as the result of successive transformations in a radioactive series; may be radioactive or stable.

Decommissioning—The process of removing a facility or area from operation and decontaminating and/or disposing of it or placing it in a condition of standby with appropriate controls and safeguards.

Decontamination—The selective removal of radioactive material from a surface or from within another material.

Decrepitation—Evaporation and explosion of water contained in salt.

Diapir—A geologic flow structure, either a dome or an anticline, in which overlying rocks have been ruptured by the upward flow of a plastic core material such as salt.

Diapirism—The disruption of a geological layer by an underlying layer; intrusion of one layer into the second.

Disposal—Confinement of radioactive waste in such a way that its separation from the biosphere is expected to be permanent, with no need for further surveillance.

Dissolution front—The boundary of a geologic region within which salt is dissolving.

Dose—A general form denoting the quality of radiation or energy absorbed. For special purposes, it must be appropriately qualified. If unqualified, it refers to absorbed dose.

Dose limit (maximum permissible dose)—The dose of ionizing radiation established by competent authorities as an amount below which there is no reasonable expectation of risk to human health and that at the same time is somewhat below the lowest level at which a definite hazard is believed to exist.

Dose risk—The quantitative risk of contracting cancer as a result of a given radiation dose.

Electromagnetic radiation—Radiation consisting of associated and interacting electric and magnetic waves that travel at the speed of light, such as visible light, radiowaves, microwaves, gamma rays, and x-rays.

Electron—Charged particle that orbits around an atomic nucleus; see *Beta particle.*

Enrichment, isotopic—A process by which the relative abundance of the isotopes of a given element are altered, thus producing a form of the element that has been enriched in one particular isotope.

Entombment—Decommissioning of a nuclear facility by sealing in concrete.

Exponential decay—In the context of this report, the mathematical function that describes radioactive decay—that is, after one half-life, half of the original material remains; after two half-lives, one-quarter; after three half-lives, one-eighth; and so on.

Exposure—A measure of the ionization produced in air by x or gamma radiation. Acute exposure generally refers to a high level of exposure of short duration; chronic exposure is lower level exposure of long duration.

Fissile—Describes a nuclide that undergoes fission on absorption of neutrons of any energy.

Fission—The splitting of a heavy nucleus into two approximately equal parts (which are nuclei of lighter elements), accompanied by the release of a relatively large amount of energy and generally one or more neutrons. Fission can occur spontaneously, but usually is caused by nuclear absorption of gamma rays, neutrons, or other particles.

Fissionable—Describes a nuclide that undergoes fission on absorption of a neutron over some threshold energy.

Fission products—The nuclei (fission fragments) formed by the fission of heavy elements, plus the nuclides formed by the radioactive decay of fission fragments.

Food chain—The pathways by which any material (such as radioactive material from fallout) passes from the first absorbing organism through plants and animals to man.

Fuel cycle—The complete series of steps involved in supplying fuel for nuclear reactors. It includes mining, refining, the original fabrication of fuel elements, their use in a reactor, and management of spent fuel and radioactive wastes. May also include chemical reprocessing to recover the fissionable material remaining in the spent fuel.

Gamma rays—Short wavelength electromagnetic radiation, often of high energy, emitted from the nucleus.

Gaseous centrifuge process—A method of isotopic separation in which heavier gaseous atoms or molecules are separated from lighter ones by centrifugal force.

Gaseous diffusion—A method of isotopic separation based on the fact that gas atoms or molecules with different masses will diffuse through a porous barrier (or membrane) at different rates. This method is used by DOE to separate U-235 from U-238; it requires large plants and enormous amounts of electrical power.

Genetic mutations—Radiation-induced chromosomal changes that occur in the gonads and can be passed on to subsequent, unexposed generations.

Geologic isolation—The disposal of radioactive wastes deep beneath the earth's surface.

Gigawatt (GWe)—Billions of watts of electricity generated. 1 GWe = 1,000 MWe.

Gigawatt year (GWe yr)—One billion watts generated constantly throughout one year.

Groundwater—Water that exists or flows below the earth's surface (within the zone of saturation).

Half-life—The time in which half the atoms of a particular radioactive substance disintegrate to another nuclear form. May range from millionths of a second to billions of years. After a period of time equal to ten half-lives, the radioactivity level of a radionuclide has decreased to 0.1 percent of its original value; after twenty half-lives, to less than one-millionth of its original value.

High level waste—See *Radioactive waste*.

Hot spot—A surface area of higher than average radioactivity.

Hydrogeology—The science of phenomena and of the distribution of the waters of the earth.

Hydrology—See *Hydrogeology*.

Impermeable—Resistant to penetration by water.

Inclusion, brine—A small opening in a rock mass (salt) containing brine; also, the brine included in such an opening. Some gas is often also present.

Ingestion hazard index—That quantity of water required to dilute a given amount of radioactive material to maximum permissible concentration levels for water.

Inhalation hazard index—That quantity of air required to dilute a given amount of radioactive material to maximum permissible concentration levels for air.

Interim storage—Storage operations for which (1) monitoring and human control are provided and (2) subsequent action involving treatment, transportation, or final disposition is expected.

In situ—In the natural or original position; used to refer to inplace experiments at a repository site.

Institution—In this instance, those governmental agencies responsible for the public weal.

Iodine—Element 53. Iodine–129 and 131 are radioactive fission products with half-lives of sixteen million years and eight days, respectively. Iodine has a special affinity for the thyroid gland.

Ion—An atom or molecule that has lost or gained one or more electrons. By this process of ionization it becomes electrically charged.

Ion exchange—A chemical process involving the reversible interchange of various ions between a solution and a solid material. It is used to separate and purify chemicals, such as fission products or rare earths in solution. This process also takes place with many minerals found in nature and with ions in solution, such as groundwater.

Ionic retention—See *sorption*.

Ionizing radiation—Any radiation displacing electrons from atoms or molecules, thereby producing ions. Ionizing radiation may produce severe skin or tissue damage.

Irradiation—Deliberate exposure of a substance to radioactivity; in this instance, the exposure of reactor fuel to neutrons produced in the fission process.

Isolation—An alternative term for disposal in which the waste is emplaced without provision for recovery.

Isotope—One of two or more atoms with the same atomic number (the same chemical element) but with different atomic weights, i.e., the same number of protons but different numbers of neutrons. Chemical properties of isotopes are usually identical, while radioactive properties may vary enormously.

Kiloton—A nuclear explosion whose total energy release is equal to 1,000 tons of TNT.

Krypton—Element 36. A noble gas, it does not interact chemically with any other element. Krypton-85 is a radioactive fission product contained in spent reactor fuel, which is released during reprocessing. The half-life of this isotope is eleven years.

Laser enrichment—Enrichment through selective laser-induced excitation of one isotope preferentially to another.

Latent period—The period or state of seeming inactivity between the time of exposure of tissue to an acute radiation dose and the onset of the final stage of radiation sickness.

Leaching—The process of extracting a soluble component from a solid by the percolation of a solvent (e.g., water) through the solid.

Linear hypothesis—The assumption that a dose-effect curve derived from data in the high dose and high dose rate ranges may be extrapolated through the low dose and low dose range to zero, implying that, theoretically, any amount of radiation will cause some damage.

Liquid waste—Liquid radioactive wastes produced by one of the waste streams of the nuclear fuel cycle. May be very dilute with low activity or highly concentrated with high activity. See *High level waste* and *Low level liquid waste.*

Low level waste—Solid radioactive waste consisting of contaminated machinery, gloves, aprons, tissues, paper, etc., with an average activity of less than one curie per cubic foot of material or less than ten billionths of a curie of alpha activity per gram.

Low level liquid waste—Fluid materials that are contaminated by less than fifty billionths per liter of mixed fission products.

Matrix, waste—The material in which radioactive nuclear waste is encapsulated, usually a material such as borosilicate glass.

Maximum permissible concentration (MPC)—The amount of radioactive material in air, water, or food that might be expected to result in a *maximum permissible radiation dose* to persons consuming them at a standard rate of intake and from which a person, if continuously exposed, should not (in the light of present knowledge) sustain appreciable body damage.

Medium, disposal—The rock or mineral body into which radioactive waste canisters are emplaced.

Mill tailings—The fine, gray sand, with 700 trillionths of a curie per gram of radioactivity, left over after the refining of uranium ore to obtain uranium oxides.

Megawatt (MWe)—One million watts of electricity.

Megawatt year (MWe yr)—One million watts of electricity generated constantly throughout one year.

Micro—A prefix that divides a basic unit by one million.

Milli—A prefix that divided a basic unit by one thousand.

Mobility—The ability of radionuclides to move through food chains in the environment.

Morbidity—The condition of being diseased or the ratio of sick to well persons in a community.

Mothballing—Decommissioning a nuclear facility by placing under lock and key.

Nano—A prefix that divides a basic unit by one billion.

Neutrons—A subatomic particle with zero electric charge and with a mass nearly that of a hydrogen atom. Low energy neutrons can induce fission in fissionable materials.

Nonhigh level waste—See *Low level waste.*

Nuclear energy—See *Atomic energy.*

Nuclear fuel cycle—See *Fuel cycle.*

Nuclear reactor—A device in which a controlled nuclear chain reaction is maintained, either for the purposes of experimentation, production of weapons grade fissionable material, or generation of electrical power.

Nucleus—The center of an atom in which protons and neutrons are located.

Nuclide—A species of atom having a specific mass, atomic number, and nuclear energy state. These factors determine the other properties of the element, including its radioactivity.

NWTS—National Waste Terminal Storage Program. DOE program for locating suitable repository sites and constructing a commercial HLW repository.

Organ-seeking—A radionuclide with an affinity for particular biologic organs. The affinity may be the result of the organ's inability to distinguish between isotopes, as in the case of radioactive and nonradioactive iodine, or because of chemical similarity, as in the case of calcium and strontium.

Overpack—Secondary (or additional) external containment for packaged nuclear waste.

Parent—A radionuclide that upon disintegration yields a specified nuclide, the *daughter*, either directly or as a later member of a radioactive decay series.

Partition—To separate an element from others; the separation of uranium and plutonium from the spent fuel solution through the reprocessing solvent extraction process.

Partitioning—The process of separating liquid waste into two or more fractions, specifically with reference to the removal of certain radioisotopes from the waste in order to facilitate subsequent waste storage and disposal.

Pathway—See *Food chain*.

Permeability—A quantity that describes the rate at which water flows through an aquifer.

Plutonium—Element 94. A highly toxic actinide, plutonium has isotopes with a wide range of half-lives. Of interest are plutonium-238, with a half-life of 86 years, and plutonium-239, with a half-life of 24,400 years. Extremely small quantities of plutonium are capable of inducing lung cancer.

Proliferation, nuclear—The acquisition of nuclear weapons by presently non-nuclear states.

Proton—A stable charged particle found in the atomic nucleus; a hydrogen nucleus.

Purex—Abbreviation for plutonium-uranium extraction.

Purex process—An aqueous reprocessing method that consists of the following steps: (1) solvent extraction that affects the partition of uranium and plutonium from the fission products in the spent fuel solution; (2) purification of uranium and plutonium; (3) conversion of uranium to UF_6 and plutonium to PuO_2.

Rad (Radiation Absorbed Dose)—The basic unit of absorbed dose of ionizing radiation. A dose of one rad means the absorption of 100 ergs of radiation energy per gram of absorbing material.

Radiation—The emission and propagation of energy through material or space by means of electromagnetic disturbances, which display both wavelike and particlelike behavior; the energy so propagated. The term has been extended to include nuclear radiation, such as alpha and beta particles, neutrons, gamma rays, and x-rays.

Radiation sickness—The complex of symptoms resulting from excessive exposure of the whole body to ionizing radiation. Symptoms include nausea, vomiting, diarrhea, loss of hair, hemorrhage, inflammation of the mouth and throat, and general loss of energy. In severe cases, where exposure has been relatively large, death may occur within two to four weeks. Those who survive six weeks after receiving a single large dose may be expected to recover.

Radioactivity—The spontaneous decay or disintegration of an unstable atomic nucleus, usually accompanied by the emission of ionizing radiation. The word "radioactivity" is often used to refer to radioactive materials or radioactive nuclides, but strictly speaking, this usage is not correct. Radioactivity is a process, not a material substance.

Radioactive waste—Useless by-products produced in the nuclear fuel cycle. Radioactive wastes are generally divided into several different categories:

1. Low level wastes may be solid or liquid and consist of garbage generated by the handling of radioactive substances or of the dilute waste streams produced in the fuel cycle.
2. High level wastes generally refer to those highly radioactive liquids produced by the reprocessing of spent fuel and may contain thousands of curies of activity per gallon. Also may be used in reference to spent reactor fuel.
3. Transuranic-contaminated wastes are those low level wastes that contain more than ten nanocuries of transuranic (alpha) nuclide radioactivity per gram. Such waste is specially handled due to the long-lived nature of some of the transuranic nuclides.

Radiohydrology—Same as hydrogeology, but utilizing the presence of natural radionuclides or manmade radionuclides injected into groundwater systems as a study aid.

Radioisotope—A radioactive isotope. An unstable isotope of an element that decays or disintegrates spontaneously, emitting radiation. More than 1,300 natural and artificial radioisotopes have been identified.

Radiological—Pertaining to radiation and health.

Radionuclide—A nuclide that is radioactive.

Radium—Element 88. A radioactive metallic element whose most common isotope, Ra-226, has a half-life of 1,600 years. Radium-226 is the parent of radon gas.

Radon—Element 86. A radioactive gas, with a half-life of 3.8 days, that is produced by the decay of one of the daughters of radium. Radon is hazardous in unventilated areas because it can build up to high concentrations and, if inhaled for long periods of time, may induce lung cancer.

Radwaste—Contraction of "radioactive waste."

Raffinate, aqueous—Liquid left in the solvent extraction system, from which the uranium has been extracted by contact with an immiscible (two liquids that do not mix and form more than one phase when brought together) organic solvent.

Reactor—See *Nuclear reactor.*

Reactor core—That part of a nuclear reactor containing the fuel rods in which the controlled chain reaction takes place.

Reactor year—Operation of a nuclear reactor for a period of one year.

Radiation concentration guide (RCG)—A federally promulgated standard that gives the allowable limit for radionuclides in air and water intended for public use (same as MPC).

Relative biological effectiveness (RBE)—A factor used to compare the biological effectiveness of absorbed radiation doses (i.e., rads) due to different types of ionizing radiation.

Rem (roentgen equivalent man)—An unofficial, but widely used, unit of measure for the dose of ionizing radiation that gives the same biological effect as one roentgen of x-rays. One rem equals approximately one rad for x, gamma, or beta radiation, and one-tenth rad for alpha radiation.

Repository—A term generally applied to a facility for the disposal of radioactive wastes, particularly high level waste and spent fuel.

Reprocessing—That chemical process by which unfissioned uranium-235 and plutonium-239 are removed from spent reactor fuel for utilization in new fuel rod fabrication.

Retrievability—A term used to designate the ability to remove high level waste or spent fuel from a repository after emplacement, thereby allowing rectification of unforseen errors.

RSSF (Retrievable Surface Storage Facility)—A surface facility for the interim storage of high level reprocessing wastes for periods of up to one hundred years, RSSF would consist of a field of mausolea into which the high level waste canisters would be emplaced.

Risk—The product of probability and consequence—e.g., the radioactive risk of a scenario is the population dose resulting from that scenario multiplied by the probability that the scenario will actually occur.

Salt bed—A layer or layers of salt laid down during earlier geological epochs by the evaporation of a prehistoric sea.

Salt cake—The solid residue resulting from a concentration of high level waste in underground waste storage tanks.

Scientific feasibility—That condition of an idea to which no scientific objections can be found.

Seepage pond—An artificial body of surface water formed by discharge of liquid waste (also Holding pond).

Seismic—Of, subject to, or caused by an earthquake.

Shale—A sedimentary (layered) rock formed by the consolidation of clay, mud, or silt, generally highly resistant to the passage of water (highly impermeable).

Solid waste—Either solid radioactive material or solid objects that contain radioactive material or bear radioactive surface contamination.

Soluble—Able to be dissolved in a solvent—e.g., salt is highly soluble in water.

Somatic effects—Effects of radiation limited to the exposed individual, as distinguished from genetic effects, such as radiation sickness, leukemia, or cancer.

Sorption—In chemistry and geochemistry, the general term for the retention of one substance by another by close range chemical or physical forces. Absorption takes place within the pores of a granular or fibrous material. Adsorption takes place largely at the surface of a material or its particles.

Spent fuel—Irradiated reactor fuel that has been removed from the reactor core. In the case of power reactors, this means that the uranium-235 content has dropped below 1 percent.

Spent fuel pool—A pool of water, not unlike a swimming pool, located adjacent to the reactor building, into which freshly discharged spent fuel is placed, in order to allow the radioactivity and thermal levels to decay to manageable levels (see *Swimming Pool*).

SURFF (Spent Unreprocessed Fuel Facility)—A surface storage facility similar to *RSSF*, except that it would contain unreprocessed spent fuel.

Storage—Temporary or interim confinement of high level waste of spent fuel, in the expectation that a disposal method will be developed. All high level waste and spent fuel is presently in storage.

Strontium—Element 38. Strontium-90, with a half-life of twenty-eight years, is chemically similar to calcium and tends to accumulate in place of calcium in biologic systems.

Swimming pool—Jargon for spent fuel pool.

Synthetic minerals—A solid waste form, 20-80 percent radioactive waste by weight, in which particular minerals are chemically bonded to particular waste atoms to produce new, stable mineral species.

Technical feasibility—That condition of an idea to which no technical objections can be found and that can be technically demonstrated.

Tectonic—Of, pertaining to, or designating the rock structures resulting from deformation of the earth's crust.

Threshold hypothesis—A radiation dose consequence hypothesis that holds that biological radiation effects will occur only above some minimum dose.

Transmutation—The conversion of one nuclide to another, whether by neutron capture, as in the case of U-238 to Pu-239, or by radioactive decay, as in the case of many fission products.

Transuranic nuclide—A nuclide having an atomic number greater than that of uranium.

Transuranic-contaminated waste (TRU waste)—Low level radioactive waste containing more than ten nanocuries of transuranic (alpha) contamination per gram.

Tritium—A radioactive isotope of hydrogen with two neutrons and one proton in the nucleus; half-life of 12.3 years.

Tuff—A rock composed of the finer kinds of volcanic detritus, usually more or less stratified and in various states of consolidation.

Unsaturated zone—In hydrology, a distinct underground region that is not saturated with water.

Uranium—Element 92. Uranium contains two principal isotopes, uraniums-235 and 238. The former is fissionable and is the source of energy in a uranium-fueled nuclear reactor. The latter can be converted, by neutron capture, to plutonium-239, which is also fissionable. Natural uranium is 0.7 percent U-235.

Uranium hexafluoride—A volatile compound of uranium and fluoride. UF_6 gas is the process fluid in the gaseous diffusion process.

Vitrification—The formation of glassy or noncrystalline material out of nuclear wastes, when subjected to temperatures between 950°C and 1,150°C.

WIPP (Waste Isolation Pilot Plant)—The Department of Energy's underground disposal site for low level and TRU defense wastes and possibly trial disposal of 1,000 spent fuel assemblies, WIPP is to be located in a bedded salt formation in southeastern New Mexico at a depth of 2,100 feet.

X-rays—Penetrating electromagnetic radiation whose wavelengths are shorter than those of visible light, produced outside of the nucleus. X-rays are generally, but not always, less energetic than gamma rays.

Yellowcake—A semirefined uranium oxide extracted from raw uranium ore.

References for this Glossary: NAS, 1978; BEIR Report, 1972; Glasstone and Dolan, 1977; JPL, 1977; DOE 1979a; DOE, 1979b.

References

AEC. 1971. Atomic Energy Clearing House Reports. Washington, D.C.: U.S. Atomic Energy Commission. October 4.

_____. 1972. *Contaminated Soil Removal Facility, Richland Washington.* WASH-1520. Washington, D.C.: U.S. Atomic Energy Commission, April.

_____. 1973. *The Safety of Nuclear Power Reactors (Light Water-Cooled) and Related Facilities.* WASH-1250. Washington, D.C.: U.S. Atomic Energy Commission, July.

_____. 1974a. *Improved Control of Radioactive Waste at Hanford.* WASH-1315. Washington, D.C.: U.S. Atomic Energy Commission, Directorate of Licensing, June.

_____. 1974b. *Waste Management Operations, Hanford Reservation, Richland, Washington.* WASH-1538. Vol. 1. Washington, D.C.: U.S. Atomic Energy Commission, Draft Environmental statement, September.

Anderson, D.R.; C.D. Hollister; and D.M. Talbert. 1976. *Report to the Radioactive Waste Management Committee on the First International Workshop on Seabed Disposal of High Level Wastes, Woods Hole, MA., February 16–20, 1976.* SAND76-0224. Albuquerque, N.M.: Sandia Laboratories, April.

Anderson, R.Y. 1978. "Report to Sandia Laboratories on Deep Dissolution of Salt, Northern Delaware Basin, New Mexico." January. Mimeo.

APS. 1977. *Report to the American Physical Society by the Study Group on Nuclear Fuel Cycles and Waste Management.* New York: American Physical Society. Reprinted in *Rev. Mod. Phys.* 50, Part II (January 1978).

Ayres, R.W. 1975. "Policing Plutonium: The Civil Liberties Fallout." *Harvard Civil Rights Civil Liberties Law Review* 10, no. 2 (Spring): 369.

Bair, W.J., and R.C. Thompson. 1974. "Plutonium: Biomedical Research." *Science* 183 (February 22): 715.

Barnaby, W. 1978. "Nuclear Waste Problem Solved, Claims Sweden's Nuclear Industry." *Nature* 274 (July 6): 6.

BEIR Report. 1972. *The Effects on Populations of Exposure to Low Levels of Ionizing Radiation.* Report of the Advisory Committee on the Biological Effects of Ionizing Radiation. Washington, D.C.: National Academy of Sciences, November.

Bell, M.J. 1973. *ORIGEN, the ORNL Isotope Generation and Depletion Code.* ORNL-4628. Oak Ridge, Tenn.: Oak Ridge National Laboratory.

Bugliarello, G., and F.C. Gunther. 1974. *Computer Systems and Water Resources.* New York: Elsevier Science Publishing Company.

Bowen, V.T. 1974. "Transuranic Elements and Nuclear Wastes." *Oceanus* 18 (Fall): 48.

Bredehoeft, J.D., et al. 1978. *Geological Disposal of High-Level Radioactive Wastes—Earth Science Perspectives.* U.S. Geological Survey Circular 779. Reston, Va.: U.S. Geological Survey, May.

Bureau of Mines. 1977. *Valuation of Potash Occurrences Within the Waste Isolation Pilot Plant Site in Southeastern New Mexico.* ALO-18. Denver: U.S. Bureau of Mines, November.

Carter, L.J. 1977. "Radioactive Wastes: Some Urgent Unfinished Business." *Science* 195 (February 18): 661.

_____. 1978a. "Nuclear Wastes: Geologic Disposal Seen as Weak." *Science* 200 (June 8): 1135.

_____. 1978b. "Uranium Mill Tailings: Congress Addresses a Long-Neglected Problem." *Science* 202 (October 13): 191.

CERCDC. 1978a. *Status of Nuclear Fuel Reprocessing, Spent Fuel Storage, and High Level Waste Disposal.* Draft Report. Report by the Nuclear Fuel Cycle Committee of the California Energy Resources Conservation and Development Commission. Sacramento, Calif., January 11.

_____. 1978b. *A Review of the KBS Reports on Spent Nuclear Fuel Handling and High Level Waste Storage.* Report by the Nuclear Fuel Cycle Committee of the California Energy Resources Conservation and Development Commission. Sacramento, Calif., June.

Chapman, N.; D. Gray; and J. Mather. 1978. "Nuclear Waste Disposal: The Geological Aspects." *New Scientist*, April 27, p. 225.

Cohen B.L. 1977a. "High-level Radioactive Waste From Light-Water Reactors." *Rev. Mod. Phys.* 49 (January): 1.

_____. 1977b. "The Disposal of Radioactive Wastes from Fission Reactors." *Scientific American* 236, no. 6 (June): 21.

Cowan, G.A. 1976. "A Natural Fission Reactor." *Scientific American* 235, no. 1 (July): 36.

Dawson, P.R., and J.R. Tillerson. 1978. *Nuclear Waste Canister Thermally Induced Motion.* SAND 78-0566. Albuquerque, N.M.: Sandia Laboratories, June.

DeBuchannane, G. 1978. Statement Before the Subcommittee on Science, Technology, and Space of the Senate Committee on Commerce, Science, and Transportation. March 31. Reston, Va.: U.S. Geological Survey, Division of Radiohydrology.

DeBuchannane, G., and W. Wood. 1978. Private conversation with Drs. G. DeBuchannane and W. Wood, U.S. Geological Survey Office of Radiohydrology, Reston, Va., June 1.

Deese, D.A. 1978. *Nuclear Power and Radioactive Waste: A Sub-Seabed Disposal Option?* Lexington, Mass.: Lexington Books.

DeMarsily, G., et al. 1977. "Nuclear Waste Disposal: Can the Geologist Guarantee Isolation?" *Science* 197 (August 5): 519.

Dickey, B.R.; G.W. Hogg; and J.R. Berreth. 1978. "Calcine Production and Management." In L.A. Casey, ed., *Proceedings of the Conference on High-Level Radioactive Solid Waste Forms* (Denver, Colorado, December 19-21, 1978), NUREG/CP-0005. Washington, D.C.: Nuclear Regulatory Commission, Office of Nuclear Materials Safety and Safeguards.

DOD. 1975. *Emergency War Surgery.* Washington, D.C.: U.S. Department of Defense.

DOE. 1978a. *Report of Task Force for Review of Nuclear Waste Management.* Washington, D.C.: U.S. Department of Energy, DOE/ER-004/D, February.

_____. 1978b. *Storage of U.S. Spent Power Reactor Fuel.* Draft Environmental Impact Statement. Washington, D.C.: U.S. Department of Energy, DOE/EIS-0015/D, August.

_____. 1979a. *Management of Commercially Generated Radioactive Waste.* Draft Environmental Impact Statement. Washington, D.C.: U.S. Department of Energy, DOE/EIS-0046/D, August.

_____. 1979b. *Waste Isolation Pilot Plant.* Draft Environmental Impact Statement. Washington, D.C.: U.S. Department of Energy, DOE/EIS-0026/D, April.

_____. 1979c. *Nuclear Reactors Built, Being Built, or Planned in the United States as of Dec. 31, 1978.* TID-8200-R39. Oak Ridge, Tenn.: DOE Technical Information Center, March.

EPA. 1975. *Preliminary Data on the Occurrence of Transuranium Nuclides in the Environment at the Waste Burial Site Maxey Flats, Kentucky.* Washington, D.C.: U.S. Environmental Protection Agency, Office of Radiation Programs, EPA-520/3-74-021, February.

_____. 1978a. *State of Geological Knowledge Regarding Potential Transport of High-Level Radioactive Waste From Deep Continental Repositories.* Report of an Ad Hoc Panel of Earth Scientists. Washington, D.C.: U.S. Environmental Protection Agency, Office of Radiation Programs, EPA/520/4-78-004.

_____. 1978b. *Technical Support for Radiation Standards for High-Level Radioactive Waste Management.* "Subtask Report D: Assessment of Accidental Pathways." Contract No. 68-01-4470. To: Office of Radiation Programs, U.S. Environmental Protection Agency. By: Arthur D. Little, Inc., Cambridge, Mass. February. Draft.

ERDA. 1976. *Alternatives for Managing Wastes from Reactors and Post-Fission Operations in the LWR Fuel Cycle. Volume 4: Alternatives for Waste Isolation and Disposal.* Washington, D.C.: U.S. Energy Research and Development Administration, ERDA 76-43, May.

Fix, J.J., et al. 1977. *Environmental Surveillance at Hanford for CY-1976.* Richland, Wash.: Battelle Northwest Laboratories, BNWL-2142. (Also see previous years.)

Ford-MITRE. 1977. *Nuclear Power Issues and Choices.* Report of the Nuclear Energy Policy Study Group. Funded by the Ford Foundation; administered by the MITRE Corporation. Cambridge, Mass.: Ballinger Publishing Co.

Gillette, R. 1973. "Radiation Spill at Hanford: The Anatomy of an Accident." *Science* 181 (August 24): 728.

Glasstone, S., and P.J. Dolan. 1977. *The Effects of Nuclear Weapons.* 3rd ed. Washington, D.C.: U.S. Department of Defense and Department of Energy.

Grimwood, P., and G. Webb. 1977. "Can Nuclear Wastes be Buried at Sea?" *New Scientist*, March 24, p. 709.

GSF. 1973. *On the Safety of Disposing of Radioactive Wastes in the Asse Salt Mine.* Munich: Gesellschaft fur Strahlen-und Umweltforschung mbH.

Hambleton, W.W. 1972. "The Unsolved Problem of Nuclear Wastes." *Technology Review*, March-April, p. 15.

Heath, C. 1978. Private conversation with Dr. C. Heath, U.S. Department of Energy, Germantown, Maryland, June 1.

HEW. 1970. *Radiological Health Handbook.* Washington, D.C.: U.S. Department of Health, Education, and Welfare; Public Health Service. Revised edition.

_____. 1979. *Report of the Interagency Task Force on the Health Effects of Ionizing Radiation.* Washington, D.C.: Department of Health, Education, and Welfare. June.

Hollocher, T.C. 1975. "Storage and Disposal of High Level Radioactive Wastes." In Union of Concerned Scientists, *The Nuclear Fuel Cycle*, p. 219. Cambridge, Mass.: MIT Press.

Hollocher, T.C., and J.J. MacKenzie. 1975. "Radiation Hazards Associated with Uranium Mill Operations." In Union of Concerned Scientists, *The Nuclear Fuel Cycle*, p. 41. Cambridge, Mass.: MIT Press.

House of Representatives. 1976. *Low-Level Radioactive Waste Disposal.* Hearings Before a Subcommittee of the Committee on Government Operations, House of Representatives, 94th Cong. 2d sess. Washington, D.C.: U.S. Government Printing Office.

_____. 1978. *Nuclear Power Costs.* Twenty-third Report by the Committee on Government Operations, House of Representatives, 95th Cong., 2d sess. Washington, D.C.: U.S. Government Printing Office.

Hyder, C. 1977. *On the Geological Instabilities Produced by Buried Nuclear Wastes.* Albuquerque, N.M.: Southwest Research and Information Center, June 28.

Industrial Research/Development. 1978. "Radionuclides Running Astray." "U.S./Sweden Radioactive Wastes Program Studies Granite Storage." July, p. 28.

IRG. 1978. *Report to the President by the Interagency Review Group on Nuclear Waste Management.* Washington, D.C.: Department of Energy, TID-28817, October 19. Draft.

_____. 1979. *Report to the President by the Interagency Review Group on Nuclear Waste Management.* Washington, D.C.: Department of Energy, TID-29442, March.

Johannson, T.B., and P. Steen. 1978. *Radioactive Waste from Nuclear Power Plants: Facing the Ringhals-3 Decision.* Ds I 1978:36. Stockholm: Industridepartementet.

Johnson, K.S., and S. Gonzales. 1978. *Salt Deposits in the United States and Regional Geologic Characteristics Important for Storage of Radioactive Waste.* Y/OWI/SUB-7414/1. Oak Ridge, Tenn.: Office of Waste Isolation, Union Carbide Corp.-Nuclear Division. March.

JPL. 1977. *An Analysis of the Technical Status of High-Level Waste and Spent Fuel Management Systems.* JPL 77-69. Pasadena, Calif.: Jet Propulsion Laboratory, December 1.

KBS. 1978. *Handling of Spent Nuclear Fuel and Final Storage of Vitrified High Level Reprocessing Waste, KBS Project.* Volumes: I. General; II. Geology; III. Facilities; IV. Safety Analysis; IV. Foreign Activities. Stockholm: KBS.

Kerr, R.A. 1979a. "Nuclear Waste Disposal: Alternatives to Solidification in Glass Proposed." *Science* 204 (April 20): 289.

_____. 1979b. "Geologic Disposal of Nuclear Wastes: Salt's Lead is Challenged." *Science* 204 (May 11): 603.

Krugmann, H. 1978. "West Germany's Efforts to Close the Nuclear Fuel Cycle: Strategies for Radioactive Waste Management." *Energy Research* 2 (April-June): 107.

Krugmann, H., and F. von Hippel. 1977. "Radioactive Wastes: A Compilation of U.S. Military and Civilian Inventories." *Science* 197 (August 26): 883.

Lapedes, D.N., ed. 1974. *McGraw-Hill Encyclopedia of Environmental Science.* New York: McGraw-Hill Book Company.

Lash, T.R.; J.E. Bryson; and R. Cotton. 1974. *Citizen's Guide: The National Debate on the Handling of Radioactive Wastes from Nuclear Power Plants.* Palo Alto, Calif.: Natural Resources Defense Council.

McCarthy, G.J. 1978a. Talk given at Nuclear Regulatory Commission Conference on High-Level Radioactive Solid Waste Forms. Denver, Colorado, December 19-21.

_____. 1978b. "Crystalline and Coated High-Level Forms." In L.A. Casey, ed., *Proceedings of the Conference on High-Level Radioactive Solid Waste Forms* (Denver, Colorado, December 19-21), NUREG/CP-0005. Washington, D.C.: Nuclear Regulatory Commission. Office of Nuclear Materials Safety and Safeguards.

McCarthy, G.J., et al. 1978. "Interactions Between Nuclear Waste and Surrounding Rock." *Nature* 273 (May 18): 216.

Means, J.L.; D.A. Crerar; and J.O. Duguid. 1978. "Migration of Radioactive Wastes: Radionuclide Mobilization by Complexing Agents." *Science* 200 (June 30): 1477.

Medvedev, Z. 1976, "Two Decades of Dissidence." *New Scientist*, November 4, p. 264.

_____. 1977a. "Facts Behind the Soviet Nuclear Disaster." *New Scientist*, June 30, p. 761.

_____. 1977b. "Winged Messengers of Disaster." *New Scientist*, November 10, p. 352.

Metlay, D.S. 1978. "History and Interpretation of Radioactive Waste Management in the United States." In *Essays on Issues Relevant to the Regulation of Radioactive Waste Management*, p. 2. Washington, D.C.: U.S. Nuclear Regulatory Commission, Office of Nuclear Material Safety and Safeguards, NUREG-0412. May.

Metz, W.D. 1977. "Reprocessing: How Necessary is it for the Near Term?" *Science* 196 (April 1): 43.

Metzger, H.P. 1971. "Dear Sir, Your House is Built on Radioactive Waste." *New York Times Magazine*, October 31.

_____. 1972. *The Atomic Establishment*. New York: Simon and Schuster.

Montague, P. 1979. "Questions and Answers About the WIPP Site." Albuquerque, N.M.: Department of Planning, University of New Mexico, February 13.

Morgan, K.Z. 1978. "Cancer and Low-Level Ionizing Radiation." *Bulletin of the Atomic Scientists*, September, p. 30.

NAS. 1973. Committee on Food Protection; Food and Nutrition Board. *Radionuclides in Food*. Washington, D.C.: National Academy of Sciences-National Research Council.

_____. 1976. *National Academy of Sciences News Report* 27, #3. Quoting the Report of the Panel on Land Burial; Committee on Radioactive Waste Management, Commission on Natural Resources. Washington, D.C.: National Academy of Sciences-National Research Council.

_____. 1978. *Radioactive Wastes at the Hanford Reservation, A Technical Review*. Panel on Hanford Wastes, Committee on Radioactive Waste Management, Commission on Natural Resources. Washington, D.C.: National Academy of Sciences-National Research Council.

_____. 1979. *Risks Associated with Nuclear Power: A Critical Review of the Literature*. Summary and Synthesis Chapter. Committee on Literature Survey of Risks Associated with Nuclear Power for the Committee on Science and Public Policy. Washington, D.C.: National Academy of Sciences, April.

National Geographic Society. 1973. "Wisconsin, Michigan, and the Great Lakes." Washington, D.C.: Cartographic Division, National Geographic Society, August. (supplement to *National Geographic* 144, no. 2 (August, 1973): 147A.)

NCRP. 1971. *Basic Radiation Protection Criteria*. NCRP Report No. 39. Washington, D.C.: National Council on Radiation Protection and Measurements.

Nelson, D.C., and D.D. Wodrich. 1974. *Retrievable Surface Storage Facility for Commercial High Level Waste*. Hanford, Wash.: Atlantic-Richfield Hanford Co., ARH-SA-175. April.

New Scientist. 1977: "NASA's Reputation Goes up in Flames." *New Scientist*, October 6, p. 5.

_____. 1979. "Level Best on Radiation." *New Scientist*, January 11, p. 75.

NRC. 1975. *Reactor Safety Study*. WASH-1400. Washington, D.C.: U.S. Nuclear Regulatory Commission, October.

_____. 1976a: *The Management of Radioactive Waste: Waste Partitioning as an Alternative*. Proceedings of a Nuclear Regulatory Commission Workshop.

Seattle, Washington, June 8–10. NR–CONF–001, Washington, D.C.: U.S. Nuclear Regulatory Commission.

_____ . 1976b. *Environmental Survey of the Reprocessing and Waste Management Portions of the LWR Fuel Cycle.* W.P. Bishop and F.J. Miraglia, eds. Washington, D.C.: Nuclear Regulatory Commission, Office of Nuclear Material Safety and Safeguards, NUREG–0016, October.

_____ . 1977a. *Topical Report: RWR-1^{TM} Radwaste Volume Reduction System.* Report No. EI/NNI–77–7–NP, prepared for U.S. Nuclear Regulatory Commission by Newport News Industrial Corp. June 24.

_____ . 1977b. *Determination of Performance Criteria for High-Level Solidified Nuclear Waste.* Washington, D.C.: U.S. Nuclear Regulatory Commission, NUREG–0279, July.

_____ . 1977c. *Workshops for State Review of Site Suitability Criteria For High-Level Radioactive Waste Repositories.* Washington, D.C.: U.S. Nuclear Regulatory Commission, Office of Nuclear Material Safety and Safeguards, NUREG–0353, October.

Nuclear Industry. 1974. "Reprocessing Gap Closer." vol. 21, no. 7 (July): 8.

Nucleonics Week. 1977a. "CEQ's Gus Speth Sets Off Big Washington Row Over Waste Disposal." October 6, p. 1.

_____ . 1977b. "Giant Sponges and Sea-Dumped Radwaste; EPA Investigates." October 27, p. 12.

_____ . 1978a. "Basalt Catching up with Salt for Radwaste Repository." August 10, p. 1.

_____ . 1978b. "NRC's Bradford Proposes Conditional Nuclear Moratorium." November 23, p. 8.

_____ . 1979a. "Uncertainty About Risks Seen Forcing Radiation Dose Limits Down." January 11, p. 2.

_____ . 1979b. "DOE Putting Turkey Point Spent Fuel Into Nevada Test Site." January 18, p. 1.

_____ . 1979c. "First Signs of Congressional Opposition to Throwaway Fuel Cycle." February 22, p. 1.

_____ . 1979d. "New Threats to WIPP Waste Facility Boiling Up." March 1, p. 1.

_____ . 1979e. "Carter to Make IRG Decisions in About Three Weeks; Report Due Monday." March 8, p. 2.

_____ . 1979f. "Briefs: A shortage of plumbers and welders has delayed the chemical decontamination. . . ." March 15, p. 9.

_____ . 1979g. "DOE is Weathering a Storm of Criticism from Environmentalists. . . ." March 29, p. 7.

_____ . 1979h. "WIPP Seems Doomed; DOE Talks About a 3-Year Delay to Consider Other Sites." April 26, p. 6.

_____ . 1979i. "Stand-alone AFR Can't be Built by '83, DOE Says." February 22, p. 2.

_____ . 1979j. "Carey Salt Mine at Lyons, Kansas Could Become Low-Level Waste Facility." July 5, p. 7.

_____ . 1979k. "DOE has Changed the Scope of WIPP. . . ." August 2, p. 3.

_____. 1979l. "Final IRG Report Out; WIPP Proposal Appears in Trouble." March 15, p. 5.

_____. 1979m. "Bechtel Puts TMI-2 Restoration at Four Years and Up to $400-Million. . . . " July 19, p. 10.

OSTP. 1978a. "Isolation of Radioactive Wastes in Geologic Repositories: Status of Scientific and Technological Knowledge." Washington, D.C.: Executive Office of the President, Office of Science and Technology Policy, July 3. Draft.

_____. 1978b. *Subgroup Report on Alternative Technology Strategies for the Isolation of Nuclear Waste.* Interagency Review Group on Nuclear Waste Management, Executive Office of the President, Office of Science and Technology Policy. Washington, D.C.: U.S. Department of Energy, TID-28818, October 19. Draft.

O'Toole, T. 1979. "Soviet A-Accident Termed Worse than Three Mile Island." *Washington Post*, May 25, p. A6.

OWI. 1976. *National Waste Terminal Storage Program Progress Report for Period April 1, 1975 to September 30, 1976.* Oak Ridge, Tenn.: Office of Waste Isolation, Union Carbide Corp.–Nuclear Division, Y/OWI-8, November 30.

_____. 1978. *National Waste Terminal Storage Program Progress Report for Period October 1, 1976 to September 30, 1977.* Oak Ridge, Tenn.: Office of Waste Isolation, Union Carbide Corp.–Nuclear Division, Y/OWI-9, April.

Papadopulos, S.S., and I.J. Winograd. 1974. *Storage of Low-Level Radioactive Wastes in the Ground: Hydrogeologic and Hydrochemical Factors.* Washington, D.C.: U.S. Environmental Protection Agency, EPA-520/3-74-009, Open File Report 74-344.

Price, K.R. 1971. *A Critical Review of Biological Accumulation, Discrimination, and Uptake of Radionuclides Important to Waste Management Practices. 1943-1971.* Seattle: Battelle Memorial Institute, BNWL-B-148, December.

Ringwood, A.E., et al. 1979. "Immobilization of High Level Nuclear Reactor Wastes in SYNROC." *Nature* 278 (March 15): 219.

Rochlin, G.I. 1977. "Nuclear Waste Disposal: Two Social Criteria." *Science* 195 (January 7): 27.

Rochlin, G.I., et al. 1978. "West Valley: Remnant of the AEC." *Bulletin of the Atomic Scientists*, January, p. 22.

Rotblat, J. 1978. "The Risks for Radiation Workers." Bulletin of the Atomic Scientists, September, p. 41.

Sandia. 1978. *Geological Characterization Report-WIPP.* SAND 78-1596. Albuquerque, N.M: Sandia Laboratories, August.

Schurgin, A.S., and T.C. Hollocher. 1975. "Radiation-Induced Lung Cancers Among Uranium Miners." In Union of Concerned Scientists, *The Nuclear Fuel Cycle*, p. 9. Cambridge, Mass.: MIT Press.

Seaborg, G. 1971. *AEC Authorizing Legislation, Fiscal Year 1972.* Testimony in hearings before the Joint Committee on Atomic Energy, Part III. March 9, 16, and 17, 1971. Washington, D.C.: U.S. Government Printing Office.

Severo, R. 1979. "Hearing on Nuclear Wastes is Set for Tomorrow Upstate." *New York Times.* January 12, p. B1.

Shapiro, J. 1972. *Radiation Protection*. Cambridge, Mass.: Harvard University Press.

Shaw, M. 1971. *AEC Authorizing Legislation, Fiscal Year 1972*. Testimony in hearings before the Joint Committee on Atomic Energy, Part III. March 9, 16, and 17, 1971. Washington, D.C.: U.S. Government Printing Office.

Speth, J.G. 1979. "A Mandate from the Future: Nuclear Wastes and the Public Trust." Paper presented at the annual meeting of the American Association for the Advancement of Science, January 5, 1979, Houston, Texas. Washington, D.C.: President's Council on Environmental Quality.

Stevens, W.K. 1978. "Texas Oilfield is Revived by the Increase in Prices." *New York Times*, April 24, p. A18.

Tannenbaum. J.A. 1977. "White Elephant? Big Plant to Recycle Nuclear Fuel is Hit by Delays, Cost Rises." *Wall Street Journal*, February 17, p. 1.

Trabalka, J.R., et al. 1979. "Another Prespective of the 1958 Soviet Nuclear Accident." *Nuclear Safety* 20, no. 2 (March–April): 206.

UCS. 1978. "Radioactive Wastes vs. Plutonium Reprocessing: A False Dichotomy." Cambridge, Mass.: Union of Concerned Scientists.

Varanini, Commissioner E.E., III. 1978. Statement before the Hearings of the Subcommittee on Nuclear Regulation, Senate Committee on Environment and Public Works, on Nuclear Waste Management, April 4. Sacramento, Calif.: California Energy Resources Conservation and Development Commission.

Weinberg, A.M. 1972. "Social Institutions and Nuclear Energy." *Science* 177 (July 7): 34.

White, S. 1977. "CIA Confirms Medvedev's Disaster Claim." *New Scientist*, December 1, p. 547.

Wilson, C.L. 1979. "Nuclear Energy: What Went Wrong." *Bulletin of the Atomic Scientists*, June, p. 13.

Winchester, J.W. 1978a. "Long Term Geochemical Interactions of High Active Waste with Crystalline Rock Repository Media." Paper presented to the International Symposium on Science Underlying Radioactive Waste Management, sponsored by the Materials Research Society, Boston, November 29–December 1. Tallahassee: Department of Oceanography, Florida State University.

_____. 1978b. "Geochemical Questions Concerning High Active Waste Disposal in Granite." A Statement to the Swedish Energy Commission. Tallahassee: Department of Oceanography, Florida State University, June 8.

Winograd, I.J. 1976. "Radioactive Waste Storage in the Arid Zone." *EOS* (American Geophysical Union) 57, no. 4 (April): 884.

Zeller, E.J.; D.F. Saunders; and E.E. Angino. 1973. "Putting Radioactive Wastes on Ice." *Bulletin of the Atomic Scientists*, January, p. 4.

✳

Bibliography

There are literally thousands upon thousands of studies, reports, and articles on the subject of radioactive waste management and related fields. This bibliography is intended only as a guide to the literature; it is by no means comprehensive. Because radioactive waste management is such a fluid and contentious topic, policies and programs are in a constant state of flux. Therefore, the list is weighted toward journal articles and reports published from 1977–1979, but some of the more important earlier references are also listed. The reader interested in keeping abreast of the topic should regularly consult *Science*, *New Scientist*, and *Nature*, all weekly science journals. The reader interested in delving into the technical aspects of the topic should consult the technical documents and bibliographies in those documents or the International Atomic Energy Agency's *INIS Atomindex*, which can be found in most university libraries. (For references prior to 1977, consult the AEC or ERDA *Nuclear Abstracts*.) References marked with a plus (+) are useful or important source documents. Those marked with an asterisk (*) can be obtained from National Technical Information Service, U.S. Department of Commerce, 5285 Port Royal Road, Springfield, Virginia 22161. (Be advised that paper documents from NTIS are generally quite expensive.)

The bibliography is divided into seven sections, each of which is further subdivided. The seven areas are (1) general references; (2) radiation and health; (3) the nuclear fuel cycle; (4) radioactive waste management; (5) radioactive waste disposal; (6) history; (7) foreign programs and nuclear proliferation.

GENERAL REFERENCES
ON RADIOACTIVE WASTES

+American Physical Society. *Report to the American Physical Society by the Study Group on Nuclear Fuel Cycles and Waste Management.* New York, 1977. (Reprinted in Reviews of *Modern Physics* 50 [January 1978].)

+Committee on Government Operations. U.S. House of Representatives. *Nuclear Power Costs.* 95th Cong. 2d sess. House Report 95-1090. April 26, 1978. (The so-called "Ryan Report," this report discusses the probable economic costs of radioactive waste disposal. The two volumes of testimony that back up this report also contain much useful background information.)

Comptroller General of the United States. *Nuclear Energy's Dilemma: Disposing of Hazardous Radioactive Wastes Safely.* EMD 77-41. Washington, D.C.: Comptroller General of the United States, September 9, 1977.

Lash, T.; J.E. Bryson; and R. Cotton. *Citizen's Guide: The National Debate on the Handling of Radioactive Wastes from Nuclear Power Plants.* Palo Alto: Natural Resources Defense Council, 1974. (Concise review of the federal program from the viewpoint of the environmental community in 1974.)

Nuclear Energy Policy Study Group. *Nuclear Power Issues and Choices* (Ford-MITRE study). Cambridge, Mass.: Ballinger Publishing Co., 1977. (Concise review of radioactive waste problem and potential solutions.)

Nuclear Fuel Cycle Committee of the California Energy Resources Conservation and Development Commission. *Status of Nuclear Fuel Reprocessing, Spent Fuel Storage, and High Level Waste Disposal.* Sacramento: California Energy Resources Conservation and Development Commission, January 11, 1978. Draft Report. (The California Energy Commission was assigned the task of determining the feasibility of radioactive waste disposal in fulfillment of the legislative requirement that further construction of nuclear plants in that state be contingent upon the commission's findings. This report, backed up by over twenty volumes of interviews and hearings, presents the commission's conclusions. In addition, Commissioner E.E. Varanini, III, frequently speaks on the subject of scientific and technical feasibility of radioactive waste disposal.)

Nuclear Regulatory Commission. Office of Nuclear Material Safety and Safeguards. *Essays on Issues Relevant to the Regulation of Radioactive Waste Management.* NUREG-0412. Washington, D.C.: U.S. Nuclear Regulatory Commission, May 1978. (A volume of essays on the social, institutional, and political aspects of radioactive waste management. In particular, see: D. Metlay, "History and Interpretation of Radioactive Waste Management in the United States," p. 2; I.R. Hoos, "The Credibility Issue," p. 20; I.R. Hoos, "Assessment of Methodologies for Radioactive Waste Management," p. 31; and W.P. Bishop, "Observations and Impressions on the Nature of Radioactive Waste Management Problems," p. 51.)

RADIATION AND HEALTH

General References on Radioactivity
and Radiation
Brodine, V. *Radioactive Contamination.* New York: Harcourt, Brace Jovanovich, Inc., 1975. (Environmental Issues Series, Scientists' Institute for Public Information.)

Glasstone, S., and P.J. Dolan. *The Effects of Nuclear Weapons.* 3rd ed. Washington, D.C.: U.S. Department of Defense and Department of Energy, 1977.

Taylor, F., and G. Webb. "Where Does Radiation Come From?" *New Scientist,* December 21/28, 1978, p. 992.

Radiation Pathways in the Environment
Garner, R.J. *Transfer of Radioactive Materials from the Terrestrial Environment to Animals and Man.* Cleveland: C.R.C. Press, 1972.

Oversight Hearings before the Subcommittee on Energy and the Environment of the Committee on Interior and Insular Affairs. U.S. House of Representatives. *Radiological Contamination of the Oceans.* 94th Cong., 2d sess. Serial No. 94-63. July 26-27, 1976.

*Price, K.R. *A Critical Review of Biological Accumulation, Discrimination, and Uptake of Radionuclides Important to Waste Management Practices.* Seattle: Battelle Memorial Institute, December 1971. (Radioactivity at the Hanford Reservation.)

*Reichle, D.E.; D.J. Nelson; and P.B. Dunway. *Biological Concentration and Turnover of Radionuclides in Food Chains: A Bibliography.* Oak Ridge, Tenn.: Oak Ridge National Laboratory, July 1971.

Health Effects of Plutonium
and Other Actinides
+Bair, W.J., and R.C. Thompson. "Plutonium: Biomedical Effects." *Science* 183 (February 22, 1974): 715. (Cancer induction in the lungs of beagle dogs by microscopic plutonium particulates.)

Edsall, J.T. "Toxicity of Plutonium and Some Other Actinides." *Bulletin of the Atomic Scientists,* September 1976, p. 27. (Good review of the health effects of alpha radiation, radiation standards, implications in a plutonium economy.)

Gillette, R. "Plutonium (I): Questions of Health in a New Industry." *Science* 185 (September 20, 1974): 1027; and "Plutonium (II): Watching and Waiting for Adverse Effects." *Science* 185 (September 27, 1974): 1140. (Discusses health experience of workers in reprocessing and plutonium plants.)

Low Level Ionizing Radiation
+Advisory Committee on the Biological Effects of Ionizing Radiations. *The Effects on Populations of Exposure to Low Levels of Ionizing Radiation* (BEIR Reports). Washington, D.C.: National Academy of Sciences–National Research Council, 1972 and 1979.

Journal articles on 1972 BEIR Report:

Bulletin of the Atomic Scientists. "The BEIR Report: Effects on Populations of Exposure to Low Levels of Ionizing Radiations." March 1973, p. 47.

Gillette, R. "Radiation Standards: The Last Word or at Least a Definitive One." *Science* 177 (December 1, 1972): 966.

Journal articles on 1979 BEIR Report:

Cookson, C. "U.S. Report Fuels Radiation Controversy." *New Scientist,* May 10, 1979, p. 427.

Dickson, D. "U.S. Academy Denies Threshold for Radiation Damage." *Nature* 279 (May 10, 1979): 90.

Marshall, E. "NAS Study on Radiation Takes the Middle Road." *Science* 204 (May 18, 1979): 711.

Department of Health, Education, and Welfare. *Report of the Interagency Task Force on Ionizing Radiation.* Final Report. Washington, D.C.: Department of Health, Education, and Welfare, June 1979. (See particularly the Subgroup Report on Biological Effects of Ionizing Radiation.)

Marx, J.L. "Low-Level Radiation: Just How Bad Is It?" *Science* 204 (April 13, 1979): 160. (Review of findings of Mancuso, Stewart, and Kneale study on Hanford radiation workers and work of Irwin Bross.)

Maugh, T.M. "Chemical Carcinogens: How Dangerous are Low Doses?" *Science* 202 (October 6, 1978): 37. (Discussion of dose-response curves and low level thresholds.)

The Low Level Radiation Controversy:

Morgan, K.Z. "Cancer and Low-Level Ionizing Radiation." *Bulletin of the Atomic Scientists,* September 1978, p. 30.

Letters from: B. Cohen; K.Z. Morgan. "What is the Misunderstanding all About?" *Bulletin of the Atomic Scientists,* February 1979, p. 53.

Morgan, K.Z. "How Dangerous is Low-Level Radiation?" *New Scientist,* April 5, 1979, p. 18.

Mole, R. "Radiation Risks—A Rejoinder." *New Scientist,* May 10, 1979, p. 440.

(Articles and replies that present both sides of the low level radiation controversy.)

Uranium Mining and Milling

Archer, V.E.; J.K. Wagoner; and F.E. Lundin. "Lung Cancer Among Uranium Miners in the United States." *Health Physics* 25 (October 1973): 351.

Archer, V.E., et al. "Respiratory Disease Mortality Among Uranium Miners." *Annals of the New York Academy of Sciences* 271 (April 1977): 280.

*Environmental Protection Agency. Office of Radiation Programs. *Potential Radiological Impact of Airborne Releases and Direct Gamma Radiation to*

Individuals Living Near Inactive Uranium Mill Tailings Piles. Washington, D.C.: U.S. Environmental Protection Agency. EPA-520/1-76-001. January 1976.

Pohl, R.O. *Nuclear Energy: Health Effects of Thorium-230.* Ithaca, N.Y.: Department of Physics, Cornell University, May 1975.

+Union of Concerned Scientists. *The Nuclear Fuel Cycle.* Cambridge, Mass.: MIT Press, 1975. (See A.S. Schurgin and T.C. Hollocher, "Radiation-Induced Lung Cancers Among Uranium Miners," p. 9; T.C. Hollocher and J.J. MacKenzie, "Radiation Hazards Associated with Uranium and Mill Operations," p. 41.)

THE NUCLEAR FUEL CYCLE

Uranium Mining and Milling

Carter, L.J. "Uranium Mill Tailings: Congress Addresses a Long-Neglected Problem." *Science* 202 (October 13, 1978): 191. (Current legal status of mill tailings and cleanup efforts.)

General Accounting Office. *The Uranium Mill Tailings Cleanup: Federal Leadership at Last?* Washington, D.C.: General Accounting Office. EMD-78-90. June 20, 1978.

Hearings before the Subcommittee on Energy and Power of the Committee on Interstate and Foreign Commerce. U.S. House of Representatives. *Uranium Mill Tailings Control Act of 1978.* 95th Cong. 2nd sess. Serial No. 95-175. June 19, 1978.

Sweet, W. "Unresolved: The Front End of Nuclear Waste Disposal." *Bulletin of the Atomic Scientists*, May 1979, p. 44.

Reprocessing and Recycling

Ayres, R.W. "Policing Plutonium: The Civil Liberties Fallout." *Harvard Civil Rights Civil Liberties Law Review* 10, no. 2 (Spring 1975): 369. (Security requirements of a plutonium-based economy.)

+Bebbington, W.P. "The Reprocessing of Nuclear Fuels." *Scientific American*, December 1976, p. 30.

+Metz, W.D. "Reprocessing: How Necessary Is It for the Near Term?" *Science* 196 (April 1, 1977): 43. (From an energy and waste management point of view, reprocessing is not necessary; also discusses American experience with reprocessing.)

Spent Fuel Storage

*Benjamin, L.N. *Spent Fuel Heatup Following Loss of Water During Storage.* NUREG/CR-0649. Washington, D.C.: U.S. Nuclear Regulatory Commission, Office of Nuclear Material Safety and Safeguards, March 1979.

*Department of Energy. "Charge For Spent Fuel Storage." Draft Environmental Impact Statement, DOE/EIS-0041-D. Washington, D.C.: U.S. Department of Energy, December 1978.

_____. "Storage of U.S. Spent Power Reactor Fuel." Draft Environmental Impact Statement, DOE/EIS-0015-D. Washington, D.C.: U.S. Department of Energy, August 1978.

MHB Technical Associates. *Spent Fuel Disposal Costs.* Washington, D.C.: Natural Resources Defense Council, August 31, 1978. (Finds costs of spent fuel disposal to greatly exceed those predicted by DOE.)

Decontamination and Decommissioning
Comptroller General of the United States. *Cleaning Up the Remains of Nuclear Facilities—A Multibillion Dollar Problem.* A Report to the Congress. Washington, D.C.: General Accounting Office, 1977.

Harwood, S., et al. *Activation Products in a Nuclear Reactor.* Buffalo, N.Y.: New York Public Interest Research Group, 1976.

Manion, W.J., and T.S. LaGuardia. *An Engineering Evaluation of Nuclear Power Reactor Decommissioning Alternatives.* AIF/NESP-009. Washington, D.C.: Atomic Industrial Forum, November 1976.

Sefcik, J.A. "Decommissioning Commercial Nuclear Reactors." *Tech. Review*, June–July 1979, p. 56.

Transportation
Lippek, H.E., with C.R. Schuller. *Legal, Institutional, and Political Issues in Transportation of Nuclear Materials at the Back End of the LWR Nuclear Fuel Cycle.* Seattle: Battelle Human Affairs Research Center, September 30, 1977.

Nuclear Regulatory Commission. *Transportation of Radioactive Material in the United States.* NUREG-0073. Washington, D.C.: U.S. Nuclear Regulatory Commission, Office of Standards Development, May 1976.

+Sandia Laboratories. *Transport of Radionuclides in Urban Environs: A Working Draft Assessment.* Generic Environmental Assessment on Transportation of Radioactive Materials Near or Through a Large Densely Populated Area. Albuquerque, N.M., May 1978.

Shapiro, F.C. "Radiation Route." *New Yorker*, November 13, 1978. (Movement of radioactive wastes through New York City.)

U.S. Department of Transportation. *A Review of the Department of Transportation Regulations for Transportation of Radioactive Materials.* Washington, D.C.: U.S. Department of Transportation, Materials Transportation Bureau, Office of Hazardous Materials, October 1977.

RADIOACTIVE WASTE MANAGEMENT

General References
+Jet Propulsion Laboratory. *An Analysis of the Technical Status of High-Level Waste and Spent Fuel Management Systems.* Pasadena: California Institute of Technology, December 1, 1977. (This report was prepared for the California Energy Commission. It provides a useful technical background to the general topic of waste management.)

Low Level Waste Disposal on Land
*Fore, C.S.; N.D. Vaughan; and J. Tappen. *Shallow Land Burial of Low-Level Radioactive Wastes. A Selected Bibliography.* ORNL/EIS-133/V 1. Oak Ridge, Tenn.: Oak Ridge National Laboratory, June, 1978.

General Accounting Office. *Improvements Needed in the Land Disposal of Radioactive Wastes—A Problem of Centuries.* RED-76-54. Washington, D.C.: General Accounting Office, January 12, 1976.

House of Representatives. *Low-Level Radioactive Waste Disposal.* Hearings before a Subcommittee of the Committee on Government Operations. 94th Cong. 2d sess. February 23, March 12, April 6, 1976. Washington, D.C.: U.S. Government Printing Office. (Good discussions on Maxey Flats and Beatty disposal site problems.)

National Academy of Sciences–National Research Council. *The Shallow Land Burial of Low-Level Radioactively Contaminated Solid Waste.* Panel on Land Burial, Committee on Radioactive Waste. Washington, D.C.: Commission on Natural Resources, 1976.

Low Level Waste Disposal at Sea

Bowen, V.T. "Transuranic Elements and Nuclear Wastes." *Oceanus* 18 (Fall 1974): 48.

*Stanley, H.G., and D.W. Kaplenek. *Bibliography on Ocean Waste Disposal.* 2nd ed. Final Report, PB-265831. Anaheim, Calif.: Interstate Electronics Corp., Environmental Engineering Division, September 1976.

Turner, B. "Nuclear Waste Drop in the Ocean." *New Scientist*, October 30, 1975, p. 290.

Radioactive Waste Chemistry

*Allen, E.J. *Criticality Analysis of Aggregations of Actinides from Commercial Nuclear Waste in Geologic Storage.* ORNL-TM-6485. Oak Ridge, Tenn.: Oak Ridge National Laboratory, June 1978.

Brookins, D.C. "Oklo Reactor Reanalyzed." *Geotimes* 23, no. 3 (1978): 27.

Cowan, G.A. "A Natural Fission Reactor." *Scientific American*, July 1976, p. 36.

*Gera, F. *Geochemical Behavior of Long-Lived Radioactive Wastes.* ORNL-TM-4481. Oak Ridge, Tenn.: Oak Ridge National Laboratory, 1975.

Means, J.L.; D.A. Crerar; and J.O. Duguid. "Migration of Radioactive Wastes: Radionuclide Mobilization by Complexing Agents." *Science* 200 (June 30, 1978): 1477.

Winchester, J.W. "Long-Term Geochemical Interactions of High Active Waste with Crystalline Rock Repository Media." Paper presented to the International Symposium on Science Underlying Radioactive Waste Management. The Materials Research Society. Boston, Mass. November 29–December 1, 1978.

_____. "Status of Geochemical Knowledge Concerning High Level Radioactive Waste Disposal in Geologic Media." Report for the Panel on Implementation Requirements of Environmental Standards Committee on Radioactive Waste Management. National Academy of Sciences. June 9, 1978.

_____. "Geochemical Questions Concerning High Active Waste Disposal in Granite." A Statement to the Swedish Energy Commission, June 8, 1978. (Professor Winchester participated in the review of the Swedish KBS study; he is a professor in the Department of Oceanography; Florida State University, Tallahassee.)

Waste Solidification and Spent
Fuel Management

*Bloemke, J.O.; D.E. Ferguson; and A.G. Groff. *Disposal of Spent Fuel.* CONF-780316-4. Oak Ridge, Tenn.: Oak Ridge National Laboratory, 1978.

*Casey, L.A., ed. *Proceedings of the Conference on High-Level Radioactive Solid Forms* (December 19-21, 1978. Denver, Colorado). NUREG-CP-0005. Washington, D.C.: U.S. Nuclear Regulatory Commission, Office of Nuclear Material Safety and Safeguards, 1979. (A technical document with a few lucid papers; the conference was particularly notable for its inclusion of environmental representatives in workshops.)

+Kerr, R.A. "Nuclear Waste Disposal: Alternatives to Solidification in Glass Proposed." *Science* 204 (April 20, 1979): 289.

+McCarthy, G.J., et al. "Interactions Between Nuclear Waste and Surrounding Rock." *Nature* 273 (May 18, 1978): 216.

+National Academy of Sciences-National Research Council. *Solidification of High-Level Radioactive Wastes.* Panel on Waste Solidification, Committee on Radioactive Waste Management. Washington, D.C., 1978.

Articles on National Academy of Sciences Waste Solidification Report:

Burnham, D. "How National Academy of Science Decided to Halt a Nuclear Waste Report is Disputed." *New York Times*, July 6, 1979.

Carter, L.J. "Academy Squabbles over Radwaste Report." *Science* 205 (July 20, 1979): 287.

Holden, C. "Panel Throws Doubt on Vitrification." *Science* 201 (August 18, 1978): 599.

(This report was not actually released in final form; objections were raised to the report's findings and consequently the National Academy decided not to publish it.)

RADIOACTIVE WASTE DISPOSAL

General References

+*Department of Energy. *Management of Commercially Generated Radioactive Waste.* Draft Environmental Impact Statement. Washington, D.C.: U.S. Department of Energy. DOE/EIS-0046-D. April 1979. (A massive, two-volume document that details the various technical, political, and institutional aspects of waste disposal technologies. Although the report may be considered to be written in an optimistic vein, it is nonetheless quite comprehensive and useful.)

+*Interagency Review Group on Nuclear Waste Management. *Subgroup Report on Alternative Technology Strategies for the Isolation of Nuclear Waste.* Washington, D.C.: Executive Office of the President Office of Science and Technology Policy, October 19, 1978. TID-28818. Draft. (A useful document that summarizes the current state of knowledge about various disposal technologies. It is not very well written, however, and is somewhat difficult to wade through.)

+Kubo, A.S., and D.J. Rose. "Disposal of Nuclear Wastes." *Science* 182 (December 21, 1975): 1205. (A good summary of disposal alternatives.)

Geologic Disposal

+*Report of an Ad Hoc Panel of Earth Scientists. *State of Geological Knowledge Regarding Potential Transport of High-Level Radioactive Waste from Deep Continental Repositories.* Washington, D.C.: U.S. Environmental Protection Agency, Office of Radiation Programs, 1978. (Although regarded as overly pessimistic by some, this report is extraordinarily candid about the gaps in information required to reasonably assure successful isolation of radioactive wastes.)

+Bredehoeft, J.D., et al. *Geologic Disposal of High-Level Radioactive Wastes— Earth-Science Perspectives.* Geological Survey Circular 779. Reston, Va.: U.S. Geologic Survey, 1978. (Although somewhat technical, this report should also be read for its review of knowledge gaps.)

+DeMarsily, G., et al. "Nuclear Waste Disposal: Can the Geologist Guarantee Isolation?" *Science* 197 (August 5, 1977): 519.

+Kerr, R.A. "Geologic Disposal of Nuclear Wastes: Salt's Lead is Challenged." *Science* 204 (May 11, 1979): 603.

Salt Disposal

Anderson, R.Y. "Report to Sandia Laboratories on Deep Dissolution of Salt, Northern Delaware Basin, New Mexico." January 1978. Photocopy. (Available from the Southwest Research and Information Center. POB. 4524. Albuquerque, N.M. 87106.)

*Clairborne, H.C., and F. Gera. *Potential Containment Failure Mechanisms and their Consequences at a Radioactive Waste Repository in Bedded Salt in New Mexico.* ORNL-TM-4639. Oak Ridge, Tenn.: Oak Ridge National Laboratory, October 1974.

Interagency Review Group on Nuclear Waste Management. *Subgroup Report on Alternative Technology Strategies for the Isolation of Nuclear Waste.* Washington, D.C.: Executive Office of the President Office of Science and Technology Policy, October 19, 1978. Appendix A.

McClain, W.C., and A.L. Bloch. "Disposal of Radioactive Waste in Bedded Salt Formations." *Nuclear Tech.* 24 (December 1974): 398.

+National Academy of Sciences-National Research Council. *The Disposal of Radioactive Waste on Land.* Washington, D.C., 1957.

_____. Committee on Radioactive Waste Management. *Disposal of Solid Radioactive Wastes in Bedded Salt Deposits.* Washington, D.C., 1970.

Pohl, R.O. "Nuclear Waste Disposal: Can Salt be the Answer?" Ithaca: Department of Physics, Cornell University, November 1978. Photocopy.

Alternative Disposal Technologies

Ice Disposal

Department of Energy. *Management of Commercially Generated Radioactive Waste.* DOE/EIS-0046-D. Washington, D.C.: U.S. Department of Energy, April 1979, p. 3.7.1.

Mathers, C. "Disposal of High Level Waste in Polar Ice Sheets." *Australian Physics* 15, no. 2 (March 1978): 24.

+Zeller, E.J., et al. "Putting Radioactive Wastes on Ice." *Bulletin of the Atomic Scientists*, January 1973, p. 4.

Seabed Disposal
Deese, D.A. *Nuclear Power and Radioactive Waste: A Sub-Seabed Disposal Option?* Lexington, Mass.: Lexington Books, 1978.

Department of Energy. *Management of Commercially Generated Radioactive Waste.* DOE/EIS-0046-D. Washington, D.C.: U.S. Department of Energy, April 1979, p. 3.6.1.

+Grimwood, P., and G. Webb. "Can Nuclear Wastes be Buried at Sea?" *New Scientist*, March 24, 1977, p. 709.

"High-Level Nuclear Wastes in the Seabed?" *Oceanus* 20, no. 1 (Winter 1977). Special issue.

+Kerr, R.A. "Geologic Disposal of Nuclear Wastes: Salt's Lead is Challenged." *Science* 204 (May 11, 1979): 603. (Discusses seabed disposal also.)

Space Disposal
Department of Energy. *Management of Commercially Generated Radioactive Waste.* DOE/EIS-0046-D. Washington, D.C.: U.S. Department of Energy, April 1979, p. 3.10.1.

Laporte, T.; D. Metlay; and P. Windham. *Space Disposal of Nuclear Wastes. Vol. 2: Socio-Political Aspects.* N-77-23914. Berkeley: Institute of Governmental Studies, University of California, 1976.

Laporte, T.; G.I. Rochlin; D. Metlay; and P. Windham. *Space Disposal of Nuclear Wastes. Vol. 1:* N-77-24924. Los Angeles: Department of Geological Sciences, University of Southern California, December 1976.

Nicholls, R.W. "Solar Nuclear Waste Disposal." *Nature* 269 (October 13, 1977): 556.

Partitioning and Transmutation
Department of Energy. *Management of Commercially Generated Radioactive Waste.* DOE/EIS-0046-D. Washington, D.C.: U.S. Department of Energy, April 1979, p. 3.9.1.

McKay, A. "Destroying Actinides in Nuclear Reactors." *Nuclear Engineering International*, January 1978, p. 40.

Nuclear Regulatory Commission, Office of Nuclear Material Safety and Safeguards. *The Management of Radioactive Waste: Waste Partitioning as an Alternative.* NR-CONF-001. Proceedings of a Nuclear Regulatory Commission Workshop. Seattle, Washington, June 8-10, 1976. (Overly optimistic assessments of the possibilities of partitioning and transmutation.)

Other Disposal Technologies
Department of Energy. *Management of Commercially Generated Radioactive Waste.* DOE/EIS-0046-D. Washington, D.C.: U.S. Department of Energy, April, 1979.

Winograd, I.J. "Radioactive Waste Storage in the Arid Zone." *EOS* 55, no. 10 (October 1974): 884.

Disposal Hazards and Risks

Cohen, B.L. "High-Level Radioactive Waste from Light Water Reactors." *Reviews of Modern Physics* 49, no. 1 (January 1977): 1.

_____ . "The Disposal of Radioactive Wastes from Fission Reactors." *Scientific American*, June 1977, p. 21.

_____ . "Environmental Hazards in Radioactive Waste Disposal." Letter to *Physics Today*, January 1976, p. 9.

(See also "Nuclear Waste Disposal." Letters to *Physics Today*, November 1976, p. 13.)

Environmental Protection Agency. *Technical Support for Radiation Standards for High-Level Radioactive Waste Management*. "Subtask Report D: Assessment of Accidental Pathways." Office of Radiation Programs. Cambridge, Mass.: Arthur D. Little, Inc. February 1978. Draft.

_____ . "Impact Assessment of High Level Wastes." Washington, D.C.: Office of Radiation Programs, 1979.

Montague, P. "Representing the Unrepresented in Radioactive Waste Management Decisions." Paper presented to the AAAS Symposium on Radioactive Waste Management. Houston, Texas, January 5, 1979. (Available from P. Montague, Center for Environmental Research and Development, University of New Mexico, Albuquerque, N.M. 87131.)

*Pigford, T.H., and J. Choi. "Effect of Fuel Cycle Alternatives on Nuclear Waste Management." Proceedings of the Symposium on Waste Management. CONF-761020. Washington, D.C.: Energy Research and Development Administration, 1976.

HISTORY OF RADIOACTIVE WASTE MANAGEMENT

General

+Metzger, H.P. *The Atomic Establishment*. New York: Simon and Schuster, 1972. (This book is a must for anyone wishing to understand the background of the current program.)

Lyons, Kansas

*Bradshaw, R.L., and W.C. McClain. *Project Salt Vault: A Demonstration of the Disposal of High-Activity Solidified Wastes in Underground Mines*. ORNL-4555. Oak Ridge, Tenn.: Oak Ridge National Laboratory, 1971.

+Hambleton, W.W. "The Unsolved Problem of Nuclear Wastes." *Tech. Review*, March–April 1972, p. 15. (Good review of the Lyons program.)

Hanford Reservation

+Gillette, R. "Radiation Spill at Hanford: The Anatomy of an Accident." *Science* 181 (August 24, 1973): 728.

West Valley, New York

+Gillette, R. " 'Transient' Nuclear Workers: A Special Case for Standards." *Science* 186 (October 11, 1974): 125.

Lester, R.K., and D.J. Rose. "The Nuclear Wastes at West Valley, New York." *Tech. Review*, May 1977, p. 20.

+Rochlin, G.I., et al. "West Valley: Remnant of the AEC." *Bulletin of the Atomic Scientists*, January 1978, p. 17.

Uranium Mill Tailings
Metzger, H.P. "Dear Sir: Your House is Built on Radioactive Waste." *New York Times Magazine*, October 31, 1971. (See also *The Atomic Establishment*.)

RSSF
Carter, L.J. "Radioactive Wastes: Some Urgent Unfinished Business." *Science* 195 (February 18, 1977): 661.

*Nelson, D.C., and D.D. Wodrich. *Retrievable Surface Storage Facility for Commercial High Level Waste*. ARH-SA-175. Hanford, Wash.: Atlantic-Richfield Hanford Company, April 1974.

Maxey Flats, Kentucky
Browning, F. "The Nuclear Wasteland." *New Times*, July 9, 1976, p. 43.

House of Representatives. Hearings before a Subcommittee of the Committee on Government Operations. *Low-Level Radioactive Waste Disposal*. 94th Cong. 2d sess. February 23, March 12, and April 6, 1976. Washington, D.C.: U.S. Government Printing Office, 1976. (Also contains testimony on Beatty, Nevada, waste disposal site.)

FOREIGN WASTE MANAGEMENT AND NUCLEAR PROLIFERATION

Canada
Dotto, L. "The Great Nuclear Debate." *Science Forum* 64, no. 4 (November–December 1978): 41.

Uffen, R.J. "Let's Go Slowly on a Nuclear Power Program Until We've Solved Waste Problems." *Science Forum* 59 (October 1977): 3.

Federal Republic of Germany

General
Krugmann, H. "West Germany's Efforts to Close the Nuclear Fuel Cycle: Strategies for Radioactive Waste Management." *Energy Research* 2 (1978): 107.

Gorleben
Barnaby, F. "An Unclear Decision for Germany." *New Scientist*, April 26, 1979, p. 226.

Danglemeyer, D. "West Germany's Nuclear Dilemma." *New Scientist*, July 13, 1978, p. 103.

Hirsch, H. "Gorleben: Winning the Battle, Losing the War?" *Nature* 279 (May 24, 1979): 283.

Hopfner, K. "Planners Backtrack After German Nuclear Inquiry." *Nature* 278 (April 26, 1979): 774.

Johansen, A. "Environmentalists Stopper Germany's Nuclear Energy." *New Scientist*, March 22, 1979, p. 934.

_____. "Expert Confusion at Nuclear Hearings." *New Scientist* April 26, 1979, p. 247.

_____. "Red Light for German Nuclear Plant." *New Scientist*, May 24, 1979, p. 621.

New Scientist. "U.S. Reviews Nuclear Waste Disposal . . . and so Does West Germany." October 26, 1978, p. 253.

Redfern, R. "West Germany: Eyeball to Eyeball over Nuclear Fuel Reprocessing." *Nature* 278 (March 22, 1979): 297.

France

"France Completes Nuclear Fuel Cycle." *New Scientist*, March 29, 1979, p. 1013.

Lewis, P. "French Press Plan for Nuclear Wastes." *New York Times*, November 20, 1978.

Lloyd, A. "Europe's Nuclear Leader Accelerates." *New Scientist*, March 1, 1979, p. 654.

Great Britain

Chapman, N.; D. Gray; and J. Mather. "Nuclear Waste Disposal: The Geological Aspects." *New Scientist*, April 27, 1978, p. 225.

Feates, F., and N. Keen. "Researching Radioactive Waste Disposal." *New Scientist*, February 16, 1978, p. 426.

Hill, Sir John. "Nuclear Waste Disposal." *Atom* (London), May 1978, p. 122.

Morris, B., and A. Marples. "See-through Solution to Nuclear Waste." *New Scientist*, July 26, 1979, p. 276.

Sweden

Abrams, N.E. "Nuclear Politics in Sweden." *Environment* 21, no. 4 (May 1979): 6.

Barnaby, W. "Nuclear Waste Problem Solved, Claims Sweden's Nuclear Industry." *Nature* 274 (July 6, 1978): 6.

Energy Commission. Department of Industry. Government of Sweden. *Disposal of High Active Nuclear Waste: A Critical Review of the Nuclear Fuel Safety (KBS) Project on Final Disposal of Vitrified High Active Nuclear Fuel Waste.* Ds I 1978: 17. Stockholm, 1978. (May be obtained from Industridepartementet, Fack S-103. 10 Stockholm, Sweden. Request English version.)

Johansson, T.B., and P. Steen. *Radioactive Waste from Nuclear Power Plants: Facing the Ringhals-3 Decision.* Stockholm: Department of Industry, (Government of Sweden), Ds I 1978: 36, 1978. (Request English version.)

KBS Reports: "Handling of Spent Nuclear Fuel and Final Storage of Vitrified High Level Reprocessing Waste, KBS Project. Volume: I. General; II. Geology; III. Facilities; IV. Safety Analysis; V. Foreign Activities. Stockholm: KBS,

1978. (KBS. Brahegatan 47, S–102. 40 Stockholm, Sweden. English version must be specifically requested.)

Nuclear Fuel Cycle Committee, California Energy Commission. *A Review of the KBS Reports on Spent Nuclear Fuel Handling and High Level Waste Storage.* Prepared for a Special Review Committee of the Government of Sweden. Sacramento, June 1978.

Switzerland

Burgisser, H., et al. *Geologische Aspekte der Endlagerung radioaktiver Abfalle in der Schweiz. (Geological Aspects of the Disposal of Radioactive Waste in Switzerland.)* Affoltern, Switzerland: Energie-Stiftung, 1979. Report No. 6. (Available from: Buch 2000. Postfach 36. 8910 Affoltern a.A. Switzerland. Price: sfr. 10. In Swiss-German with English summary.)

Danglemeyer, D. "Swiss Prepare for Nuclear Referendums." *New Scientist,* February 8, 1979, p. 366.

Milnes, G. "Swiss Go to Polls Again on Nuclear Issue." *Nature* 278 (April 5, 1979): 500.

_____ . "What the Geologists Say About Where to Bury the Waste." *Nature* 278 (April 5, 1979): 500.

USSR

General

Belitzky, B. "The Soviet Answer to Nuclear Waste." *New Scientist,* April 21, 1977, p. 128.

Nuclear Accident in the Urals:

Bishop, J.E. "Tracking Down a Russian Nuclear Accident." *Wall Street Journal,* July 31, 1979.

Cockburn, A. "The Nuclear Disaster They Didn't Want to Tell You About." *Esquire,* April 25, 1978, p. 39.

Gillette, R. "That Soviet N–Accident—What Really Happened?" *Boston Globe,* May 24, 1979.

Medvedev, Z. "Two Decades of Dissidence." *New Scientist,* November 4, 1976, p. 264.

_____ . "Facts Behind the Soviet Nuclear Disaster." *New Scientist,* June 30, 1977, p. 761.

_____ . "Winged Messengers of Disaster." *New Scientist,* November 10, 1977, p. 352.

_____ . *Nuclear Disaster in the Urals.* New York: W.W. Norton & Co., 1979.

New Scientist. "Evidence on the Urals Incident." *New Scientist,* December 23/30, 1976, p. 632.

O'Toole, T. "Soviet A-Accident Termed Worse than Three Mile Island." *Washington Post,* May 25, 1979, p. A6.

Stubbs, P. "The 20-year Secret." *New Scientist,* November 10, 1977, p. 368.

Torrey, L. "Experts Go Critical Over Soviet Nuclear Disaster." *New Scientist,* July 26, 1979, p. 267.

Trabalka, J.R., et al. "Another Prespective of the 1958 Soviet Nuclear Accident." *Nuclear Safety* 20, no. 2 (March-April): 206.

White, S. "CIA Confirms Medvedev's Disaster Claim." *New Scientist*, December 1, 1977, p. 547.

_____. "More Soviet Fallout." *New Scientist*, August 2, 1979, p. 351.

Nuclear Proliferation

Congressional Research Service, Government Affairs Committee. U.S. Senate. *Reader on Nuclear Proliferation.* Washington, D.C., December 7, 1978.

Gilinsky, V. "Plutonium, Proliferation, and Policy." *Tech. Review*, February 1977, p. 58.

Office of Technology Assessment, Congress of the United States. *Nuclear Proliferation and Safeguards.* New York: Praeger Publishers, 1977.

Walsh, J. "Fuel Reprocessing Still the Focus of U.S. Nonproliferation Policy." *Science* 201 (August 25, 1978): 692.

Willrich, M., and T.B. Taylor. *Nuclear Theft: Risks and Safeguards.* Cambridge, Mass.: Ballinger Publishing Company, 1974.

Index

Accidental radiation, 5, 20, 25-27, 200
Ad Hoc Panel of Earth Scientists, 79, 82
Allied-General Nuclear Services, 124, 141
Alpha particles, 7-8, 33
American Physical Society (APS), 77, 81
  Study Group on Nuclear Fuel Cycle and Waste Management, 61
Anticline, 149, 172
Aquaflor, 124
Arthur D. Little, 104n, 107, 187n
Atomic Energy Act, 122
Atomic Energy Commission (AEC), 19, 62-63
  in Idaho, 144, 161
  RSSF proposal, 62-63
  salt disposal study, 118-20
  and waste management, 113, 116, 122, 129, 135, 137-39, 166-67
Atomic Industrial Forum (AIF), 22n
Away From Reactor (AFR) storage, 63

Background radiation, 14, 20, 25
Barnwell, S.C. plant, 124-25, 134, 141
Basalt, for waste disposal, 75, 143-44, 157
Battelle Memorial Institute, 142n
Battelle Pacific Northwest Laboratories, 57
Beatty, Nevada site, 134-35

Belgium, 199
Beta particles, 7-8
Biologic hazards, from radioactivity, 5-6, 14-18
Birth abnormalities, 13, 15-18, 25
Borosilicate glass, 57, 59, 81-82, 195
Bradford, Peter, 168
Breccia pipe, 147-49, 172

Calcine, 56-57
Canada, 196
Cancer, 13, 15-18, 20, 25
  and waste management facility failure, 107, 109-10, 187-89, 191
Canisters, waste, 55, 56-62, 81-82. See also Storage tanks
  uncertainty in, 79-81, 87, 118, 120-21, 166
Carter, Jimmy, 45, 140, 147, 170, 201
Cell death, 13, 15-18
Central nervous system syndrome, 15
Centrifuge, 36
Ceramics matrix, 59, 81-82
Chain reaction, 9
Charged particle accelerators, 95-96
Church, Frank, 119, 144
Civilian waste, 50, 120
Cladding, 44-45
Climate, effect on waste disposal, 78, 94, 105, 107-108
Cohen, Bernard, 107-108, 187
Colorado, 135-36
Containers, waste, 55, 56-62, 166. See also Canisters, waste; Storage tanks

241

at Hanford, 114-16, 137, 187,
   189-90
low level waste, 125
in salt disposal, 118
Cost
   of Barnwell plant, 124
   of decommissioning, 51-53
   of disposal at NFS plant, 123
   of mill tailings cleanup, 137
   for rock melting, 68
   for seabed disposal, 88
   for space disposal, 90
Cribs, 129
Curie, 11, 13

DeBuchannane, George, 77-78
Decay chain, 9-10, 35
Decommissioning, 29, 33, 50-54
Decontamination, 29, 50-54
Deep ocean sediment disposal, 85-87
Deep ocean trenches, 84-85
Department of Energy (DOE).
   See U.S. Department of Energy
Deutch Report, 51-52, 61, 135, 201
Dismantling-removal, 51-53
Dose, 14
   acute, 15
   chronic, 15-18
   maximum permissible, 22, 25
   permissible levels, 19-20, 25
   -risk relationship, 107-108, 187,
   189-94
Dresden-1 reactor, 53, 124

Earthquakes, 79, 152, 172
Economic problems, 2, 38, 45-46,
   137
Energy Resource and Development
   Administration (ERDA), 62, 114,
   120, 140-42, 144, 155
Energy Resources Conservation and
   Development Commission of
   California, 158
Enrichment, and radioactive waste, 29,
   36
Entombment, 51-52. *See also* Decommissioning
Environment
   background radiation, 14, 20, 25
   limits to radioactivity in, 22-27
   and mill tailings, 35, 54
   movement of radioactivity through,
   6, 18-19
Environmental Protection Agency
   (EPA)
   and dose-risk relationship, 108, 188n
   guidelines for radioactivity in the
   environment, 22, 24

on low level waste, 33, 125
on plutonium migration at Maxey
   Flats, 132
on reprocessing wastes, 45
on RSSF, 63
study on risks of waste management
   failure, 104n, 106, 187
on waste canisters, 79
Erg, 13n
Exponential rate, 11
Exposure, 6, 13-18
   acute, 15, 20-22, 25
   biological hazards from, 5-6, 14-18
   chronic, 15-18
   measurement of, 13-14
   permissible dose levels, 19-20
   safe levels of, 6, 15
   standards and guidelines for, 19-27

Fabrication, fuel rod, 29, 32-36
Faults
   in risk analysis, 187, 191
   in WIPP site, 149, 152, 172
Federal government, and waste management, 1-3, 122, 125
   current waste management, 1-3,
   140-55, 167-68
   for high level waste, 113-21
   for low level waste, 113-14, 125-35,
   166-67
   in the past, 114-38, 166-67
   for reprocessing, 122-25
   and West Valley plant, 123
Federal Republic of Germany, 196
Fission, nuclear, 9-11
Fission products, 9
Fission reactors, for transmutation, 96
Foods, limits for radionuclides in,
   24-25
Ford-MITRE study, 18, 61
France, 195
Fuel cycle, production of radioactive
   wastes, 29-33, 53-54
Fusion reactors, for transmutation, 96

Gamma rays, 7-8
Gastrointestinal syndrome, 15
Gaseous diffusion process, 36.
   See also Enrichment
GE reprocessing plant, 124, 141
Genetic defects, 13, 15-18, 25
Geologic isolation, 65-84, 140, 145
   advantages to, 65-66
   disposal concepts, 66-68
   disposal media, 68, 75-76
   foreign programs for, 195-96,
   199-200
   human intrusion and, 156

risks of repository failure, 104-106,
  109-11
and satisfactory waste disposal,
  169-70
technical uncertainties of, 76-80
Glass matrix, 57, 59, 81-82, 166
Granite, for waste disposal, 75, 143,
  157, 166
Greenhouse effect, 94
Groundwater, in geologic isolation,
  77-78, 80, 105, 149, 172
and risk, 107-109, 187-88, 191

Half-life, 9-11
Hambleton, William, 120
Hanford Reservation, 19, 122
  basalt disposal research at, 143-44
  critical mass at, 129-32
  military wastes at, 46, 48, 97
  waste management, 109, 114-16,
    137, 158, 162, 189
Hazards. *See also* Risks
  biological, 5-6, 14-18
  health, 2, 35, 45, 187-94
  radiological, 163
  from reprocessing, 45-46
  of waste management, 2, 35, 45, 55,
    96-97, 101-11, 163, 187-94
Health
  hazards and risks, 2, 35, 45, 187-94
  and radioactivity, 5, 13, 15-18, 25
Heat loading, 82-83, 94
High level wastes, 34, 38, 44-46, 54
  disposal, 66-67, 142
  management, 114-21, 143, 162
  in repository, 104, 147
  and reprocessing, 61-62
  in salt disposal, 83-84
  solidification of, 56-59, 62
  and spent fuel storage, 46-48, 50,
    54
High level liquid waste, 66-67, 142
  health risks from, 187, 189-90
  storage of, 56-59, 62, 114-16, 187
Host rock, 108
Hot spots, 34, 125
Human error and intrusion, 107-108,
  139, 155-59
Hydraulic fracturing, 106
Hydrocarbons, 152-55. *See also*
  Natural resources
Hydrogeologic ground characteristics,
  80, 134

Ice disposal, 92-94
Idaho, 119, 144, 161
Idaho Falls plant, 48, 57, 97, 119,
  129, 162
Inclusion, brine, 59, 80-82, 119, 149

India, 200-201
Indian Point-1 reactor, 122
Ingestion hazard index, 97-101
Institutional problems, 114, 139-40,
  157-59
Interagency Review Group on Nuclear
  Waste Management (IRG), 61,
  140, 145
Interim storage, 139, 140-41, 161
Internal transition decay, 7n
International law, 87-88
Ion-exchange, 131
Ionization, 7-8
Irradiation, 29, 38. *See also* Nuclear
  reactor operation
Isotopes, 6-7

Japan, 196

Kansas Geological Survey, 119-20
KBS Project, 199
Krugmann, H., 50

Laser enrichment, 36
Latency period, 15, 88-89
Leaching, 107-109, 187-88, 191
Licensing procedures, 144-45, 147
Linear hypothesis, 20
London Convention of 1972, 87
Long-term stability
  of geologic isolation, 76, 78-79, 80
  of ice disposal, 92, 94
  partitioning and transmutation,
    95-96
  of seabed disposal, 84, 87
Los Alamos Scientific Laboratory, 76
Low level waste, 33-34, 35-36, 52,
  84
  disposal management, 125-35,
    162-63
LWRs (Light water reactors)
  decommissioning, 51-52
  fissile atom in, 9
  quantity of radioactive waste, 33, 38
  spent fuel discharge from, 48
Lyons, Kansas, 118-20, 135n, 137,
  144, 158

Manhattan Project, 125
Matrix, waste, 79, 81-82. *See also*
  Borosilicate glass
Maxey Flats; Kentucky site, 132-34,
  158
Maximum permissible concentration
  (MPC), 22, 25
Medical radiation, 5, 14, 20, 25
Medium, disposal, 68, 75-76, 169-70.
  See also Geologic isolation; Salt
  disposal

Medvedev, Zhores, 200
Metal alloy containers, 60–61, 81–82,
    114–16, 118
Military waste, 48, 50, 114, 120, 145
Milling, and radioactive waste, 29,
    35–36, 54
Mill tailings, 35, 54
    management, 135–38, 162–63
Mined vaults, 68–80, 172
    disposal media for, 68, 75–76
    technical uncertainties of, 76–80
Mining, 29, 34–35
Morris Illinois plant, 141
Mortality rate, 15, 20, 45
    and waste management failure,
    189–90, 192
Mothballing, 51

National Academy of Science, 20, 68
    and dose-risk relationship, 109, 188n
    reports on waste management, 116,
    118, 125, 129
National Aeronautics and Space
    Administration, 88, 91
National Council on Radiation Pro-
    tection and Measurements, 20
National Environmental Policy Act
    (NEPA), 159
Natural resources
    and human intrusion, 156–57
    in WIPP site, 152–55, 168, 172
Neutrons, 6–9
Nevada Test Site, 143–44
Nuclear Engineering Company
    (NECO), 132–34
Nuclear explosive devices, 95
Nuclear fuel cycle. See Fuel cycle
Nuclear Fuel Services (NFS) plant,
    122–24
Nuclear reactors
    amount of radioactivity in, 13
    nuclear reactions in, 7, 8–9, 9–11
    operation, and radioactive waste, 29,
    38, 54
    shutdown and radioactive waste, 29,
    50–54
Nuclear Regulatory Commission
    (NRC)
    guidelines for dismantling-removal, 52
    licensing, 172
    limits on radioactivity in environ-
    ment, 22
    waste management, 114, 123, 145,
    152, 173
NWTS (National Waste Terminal
    Storage program), 141–44

Oak Ridge National Laboratory, 134,
    144, 200

Office of Nuclear Waste Isolation
    (ONWI), 142n
Office of Waste Isolation (OWI), 142
Off-site storage, 141
Oklo phenomenon, 65–66
On-site storage, 62–63, 141
Orbital transfer vehicle (OTV), 88–90
Organseeking. See Radionuclides

Particle emission, 7
Partitioning, 94–96
Penetrometer, 85
Philberth, Bernard, 92
Plasticity, 83, 85
Plutonium, 107, 114, 129, 131–32,
    134
Political problems, 2, 170, 173
    and reprocessing, 38, 45
    and seabed disposal, 87–88
    and waste management, 137–38,
    139, 143, 157-58
    of WIPP, 145, 147, 172
Project Salt Vault, 81–82, 118
Proliferation, 45
Public sector, and waste, 167, 170,
    173–74
Purex process, 44–45, 124

Rad, definition of, 13, 19
Radiation, 1
    environmental limits on, 22-27
    hazard of, 1, 5–6
    properties of, 7–9
Radiation sickness, 14–15
Radioactive decay, 6–7
Radioactive waste, 162–63, 166–68.
    See also High level waste; Low
    level waste; Waste disposal;
    Waste management
    categories of, 3, 33–34
    disposal of, 2–3, 56, 65–96, 166
    isolation, 55–56
    management, 1–3, 55–56, 56–62,
    96–111, 166
    production of, 1, 3, 29–33, 53–54
    as radiological hazard, 163
    storage, 56, 62–63
    toxicity of, 99, 101
Radioactivity
    in classification of radioactive waste,
    33–34
    in defense wastes, 50
    defined, 7
    hazards of, 3, 5-6, 13–18, 25–26, 27
    in Lake Erie and Lake Ontario, 124
    limits in environment, 22-27
    longevity of, 1–2
    measurement, 11–13
    migration, 107-108, 129–35, 149

movement through environment, 6,
18-19
nature of, 3, 6-13, 25
in spent fuel, 50
types of, 6-11
Radionuclides, 9-11
concentration mechanisms, 18
environmental limits on, 22-25
and geologic isolation, 65-66
migration, 80, 107-108, 149
organ-seeking, 18-19
and waste management, 131-32
Radon, 35, 135
Raffinate, 36
Rainwater, in radioactivity migration,
132, 135
Reactor vessel, 52
Rem, 19
defined, 13
Repository
failure and consequences, 104-106,
107-11, 187, 189-90
in geologic rock, 76-80, 105, 109
in salt, 81, 83, 105-106, 109-10
in WIPP, 152, 155
Reprocessing
foreign facilities for, 195-96,
199-200
and radioactive wastes, 29, 34,
44-46, 54, 61-62
U.S. management of, 122-25, 162
Reproductive failure, 13, 15-18
Research, 75-76, 78, 171
Retrievability, and waste disposal
geologic isolation, 79-80
in ice disposal, 92, 94
rock melting and, 68
in seabed disposal, 84-85
Richland, Washington site, 135
Risk analysis, 106-11, 187-94
Rochlin, Gene, 156
Rock melting, 66-68
Rocky Flats plant, 118
Retrievable Surface Storage Facility
(RSSF), 62-63, 120, 140, 144

Salt disposal, 65, 166
bedded, 68, 80, 82, 107, 119-20,
142, 145, 187-88, 191
domed, 68
foreign programs, 196
and human intrusion, 156-57
in Kansas, 118-20
problems with, 142-43
technical uncertainties, 76, 80-84,
147-49
for WIPP, 144-45, 147-49, 168
Sandia Laboratory, 76, 144-45, 149

Savannah River plant, 48, 97, 162
Schlesinger, James, 91
Schmitt, Harrison, 91
Seabed disposal, 84-88, 196
Seaborg, Glenn, 119
Seismic. *See* Earthquakes
Shale, for waste disposal, 75
Shaw, Milton, 119
Site selection
and geologic isolation, 76-80
and salt disposal, 119-20
and satisfactory waste program,
169-70, 171-72
Social problems, 2-3, 139, 170-71,
173-74
Solidification, 56-59, 116, 166
at West Valley plant, 123
Solid waste, 162
Space disposal, 88-92
Spent fuel
interim storage, 139, 140-41
irradiation, 29, 38
management, 60-61
reprocessing and waste, 29, 34,
44-46
in salt disposal, 84
surface storage for, 62-63
and waste storage, 46-50, 54
Speth, J. Gustave, 168
Storage
interim for spent fuel, 139, 140-41,
161
surface, 156, 158, 199
technology, 62-63
temporary, 63, 110, 114-15, 163,
168
Storage tanks. *See also* Canisters,
waste
at Hanford, 114-16, 189-90
for high level waste, 34, 44-48, 54,
56, 114-16, 120-21, 189-96
at West Valley, 124
Strontium-90, 24, 109-10, 187, 190,
191-94, 200
Subduction, 85
Subsediment bedrock, 85
SURFF (Spent Unreprocessed Fuel
Facility), 62-63
Sweden, 166, 199
Synthetic rock, 59

Technical feasibility, 169-70
Technical fixes, for waste management, 139, 167-68, 170
Technical problems
with geologic isolation, 76-80
in ice disposal, 92-94
recommendations for, 171-72

and reprocessing, 45-46
with salt disposal, 80-84
in seabed disposal, 87-88
in space disposal, 90-91
in transmutation, 95-96
and waste management, 137,
139-40, 143, 171-72
of WIPP, 147-49, 152-55, 172
Technology
waste management, failure of,
106-11, 113, 166
for waste solidification, 56-59
waste storage, 62-63
Technology, waste disposal, 2, 56,
65-96, 172
geologic isolation, 65-68, 75-84
ice disposal, 92-94
partitioning and transmutation,
94-96
seabed disposal, 84-88
space disposal, 88-92
Terrorist groups, 45
Three Mile Island, 19, 53, 123, 142
Threshold hypothesis, 20
Toxicity, of radioactive wastes, 97,
101
Transmutation, 94-96
Transportation, and waste disposal, 68
Transuranic elements, 9
Transuranium-contaminated waste,
34, 44, 56, 61
in Idaho, 119, 144, 161
in salt disposal, 84, 145
Tuff, 75-76

United Kingdom, 195
U.S. Committee on Government
Operations, 52
U.S. Department of Energy (DOE),
46, 48, 63
current waste management, 139-43
past waste management, 114, 123,
125-29
on space disposal, 88, 90-91
and waste disposal, 159, 161, 173
and WIPP, 145, 147, 168
U.S. General Accounting Office, 116
U.S. Geological Survey, 77-78, 84,
116, 131, 144, 152
U.S. Public Health Service, 35
USSR, 199-200
Uranium, 29-32, 34-36
enrichment, 29, 36

Varanini, Emilio, 158-59
Vitrification, 56-57, 166, 195, 199
Volumes of waste, 50, 61, 125
Von Hippel, F., 50

Waste, from fuel fabrication, 36
Waste disposal
current, 1-3, 140-55, 167-68
geologic isolation, 65-68, 75-84
ice disposal, 92-94
institutional factors in, 114, 139-40,
157-59
interim spent fuel storage, 140-41
National Waste Terminal Storage
program, 141-44
partitioning and transmutation,
94-96
past, 113-38, 166-67
recommendations for, 3, 171-74
requirements for, 168-71
risks of, 101-11, 187-94
seabed disposal, 84-88
space disposal, 88-92
technology, 2, 56, 65-96, 172
Waste Isolation Pilot Plant project,
144-55
Waste form, 36, 56
and salt disposal, 83-84
and seabed disposal, 87
and waste management, 56-59
Waste Isolation Pilot Plant project
(WIPP), 108n, 144-55, 167-68
Waste management
current, 1-3, 140-55, 167-68
hazards and risks, 96-111, 187-94
of high level waste, 114-21
human error and intrusion in,
107-108, 139, 155-59
low level waste disposal, 125-35
mill tailings, 135-38
in past, 113-38, 166-67
recommendations for, 3, 171-74
requirements for, 168-71
reprocessing, 122-25
technology failure, 106-111, 113,
166
Waste storage, 56, 62-63
of low level waste, 34
risks of, 101, 111
of spent fuel, 29, 46-50, 54
of transuranium-contaminated
waste, 34
Waste stream, 34, 36, 44. *See also*
High level waste
Weinberg, Alvin, 55
West Valley plant, 122-24, 134,
141-42, 158, 162
Wilson, Carroll, 166
Woods Hole, 85, 124

X-rays, 7-8

Yellowcake, 35-36

✳

About the Author

Ronnie D. Lipschutz is a physicist by training. His interest in public interest science and environmental issues brought him to the Union of Concerned Scientists where he is currently the organization's research staff specialist in radioactive waste management and editor of the quarterly newsletter *Nucleus*. Other interests include arms control, science and technology policy, and energy analysis and policy. He holds a master's degree in nuclear physics from the Massachusetts Institute of Technology and physics and liberal arts degrees from the University of Texas at Austin. He hopes to begin graduate study in an energy-related field within the near future.